Tropical Saws & Export Corporation ™

Joe Nathan Brown-Inventor, Designer, Artist,
Engineering, President, Chief Executive Officer
Mailing Address: 4304 Brooks Avenue,
West Palm Beach, Fl. 33407
Cell: (561) 291-5652, E-mail: tropicalsaws@live.com

TITLE PAGE

PERIMETER SAW ™ © ®
TM-TRADEMARK   BM-BRANDMARK

DESIGN PATENT-APPLICATION

These are the CLAIMS AND SPECIFICATIONS I.

FOR THE MODELS: A, B, C, D, E, F, G, H, I, J, K, L, M, N, O,
P, Q, R, S, T, U, V, W, X, Y AND Z

TYPES 1A, 2B, 3C, 4E, AND 5F

CLASSIFICATIONS: I-1, II-2, III-3, IV-4, V-5

January 21, 2009
May 18, 2009
December 23, 2010
January 9, 2011

SM-SERVICEMARK	L-LABEL
C-COPYRIGHT	ST-STICKER PRICE
R-REGISTRATION	SP-SALES PRICE

Page 1. of 140.

Order this book online at www.trafford.com
or email orders@trafford.com

Most Trafford titles are also available at major online book retailers.

Author Credits: The copyrights held by Joe Nathan Brown are 9, along
with 8 provisional patents on new product inventions.

Printed in the United States of America.

ISBN: 978-1-4269-5523-5 (sc)

Trafford rev. 01/29/2011

 www.trafford.com

North America & international
toll-free: 1 888 232 4444 (USA & Canada)
phone: 250 383 6864 ♦ fax: 812 355 4082

Tropical Saws & Export Corporation ™

Perimeter Saw © ® Design Patent Application
Joe Nathan Brown-Inventor, Designer,
Author ,Engineering
President, Chief Executive Officer

June 29, 2009

TABLE OF CONTENTS

Tropical Saws & Export Corporation ™

Perimeter Saw © ® Design Patent Application
Joe Nathan Brown-Inventor, Designer,
Author ,Engineering
President, Chief Executive Officer

TABLE OF CONTENTS
Continued...

**The Actions Taken For Documentations:
To protect and Secure the Intellectual Property Rights**

Tropical Saws & Export Corporation ™

Perimeter Saw-Design Patent Application
Joe Nathan Brown-Inventor, Designer, Author, Engineering
President, Chief Executive Officer

THE **DESIGN PATENT**-APPLICATION, CLAIMS AND SPECIFICATIONS
OF THE DESCRIPTIVE DRAWINGS
THE TITLE OF THE INVENTION IS: **PERIMETER SAW** © ®

1. The "**Perimeter Saw**", is the title of my new product tool. It will be sure to have customers, because the title is so accurate. It is derived from some engineering and math term words, Perimeter and Saw. It tells and describes something with a boundary, an exact dimensional reference. A per-determined area and space, spoken in a sentence is the largest part of the blade; of which I Joe Nathan Brown have invented a new pipe, and tubing cutting tool. And since, the dimensions of the blades; which will be used in the saw are from 1"one inch to 48" or four feet. And the information gathered to reach this conclusion of the name and to find that it best describes it.

2. The "**Perimeter Saw**", will be the title of this claimed product and all of the models will bare the same Trademark name. As it is there are 26 models of the Perimeter Saw. Model-M is the table model, and Model-T, is the one mounted on wheels. It will be made with and without a separate motor drive to move it from place to place for the work needs. The starting blades' dimensions for the two models will be 12" inches and will increase in sizes of an inch as 13", 14", 15", 16", and so on until the largest blade is made of 48" forty eight inches in diameter.

3. The **Perimeter Saw** is also my brand name and brand mark. The service mark will be the Perimeter Saw. So that repairs, and maintenance to the saw and work done by it will be billed and assigned to the listed and registered name of the Perimeter Saw. The saw will sell greatly with this inventive title and will go into patenting, copyrights, and trademarks filing with the use and applications of the Perimeter Saw. As for mentioned the name highlights the activity in the center of the blade and its' cutting edges; which are made to specifications.

4. The saw's **Retail** price estimation is, **$149.99** to **$1,249.99** dollars.

Tropical Saws & Export Corporation ™

Perimeter Saw-Design Patent Application
Joe Nathan Brown-Inventor, Designer, Author, Engineering

THE **DESIGN PATENT**-APPLICATION, CLAIMS AND SPECIFICATION OF
THE DESCRIPTIVE DRAWINGS

THE PERIMETER SAW © ®

5. THE BACKGROUND OF THE INVENTION: is the drawings beginning date was filed on November 9, 2002 to the United States Patent & Trademarks office in the a Disclosure Document Deposit Request. It covers the first Model-X, and was the only type of each of the models that followed. For example model-X, would have been a battery operated saw in sizes based on the blades outer diameters. And the blades' outer diameters are fixed to increase 1/16" one sixteenth of an inch. So unless the saw is made with a spring adjustment gear that have four size blades; which could be used in that model, they would all be made with the same specifications.

6. There would be 1/16" X 48" of the blades for each model which are made large enough to use them all, reducing the largest blade by the 1 1/2" inch to make room for the blade holding apparatus that hold the blade in place. And all of the saws' components will increase in size to make adjustments for the larger blades.

7. This is good, because there are now 26 models of the saw to be made and each model has up to 5 types, shapes, classifications and weights. The weights would vary according to the lengths in diameters, of the blades it holds. It would vary according to the length from the top of the saw to the bottom.

8. The shapes of the saws are the most important features; which causes the models to change its' letter identifications. The classifications are mainly derived from the blade's inner and outer diameters. The number of cutting edges and the number of gears on the blades, including the shapes of them, both help to make this determination. This is determind by the length they are in fractions.

Tropical Saws & Export Corporation ™

Perimeter Saw-Design Patent Application
Joe Nathan Brown-Inventor, Designer, Author, Engineering

THE **DESIGN PATENT**-APPLICATION, CLAIMS AND SPECIFICATIONS
OF THE DESCRIPTIVE DRAWINGS

THE PERIMETER SAW © ®

9. THE BRIEF SUMMARY OF THE INVENTION: The invention is a new product. And I Joe Nathan Brown has planned to complete the Disclosure Document Deposit Request, the Provisional Patent-Patent Pending, Design Patent and the Utility Patent by December 31, 2009. This product will perform all of the requirements made in its' design and utility patent claims in usages. Having a circular round blade with the cutting edges placed evenly in the inner (Perimeter) thereby the name of the Perimeter Saw. And the gears for the circular blades are on the circumference edges of the circular blade, this arrangement of parts and pieces make for the new original idea and invention.

10. The saw has advantages and disadvantages limitations and unlimited usages and capabilities. The first advantages will be that it, the blade covers the entire circumference of the pipe, tubing, metal, wood, or rock to be cut. It will rotate at an even movement being kept in line by the Blade's Stability Circles or Blade Stability Gears. The Blade Housing Insert Attachment is designed and engineered with 1 or 2 stability circles; which are spaced over approximately $1/16^{th}$ of an inch from the stability circle on the rear of the blade. So that the order of assembly in this upper front and top section will be: B.H.I.A., the blade housing insert attachment where the blade fits into, the blade and then the blade holding apparatus. In Model-x-1 these parts are all molded together; which is then screw tightened down on the blade. The blade stability circles 1 and or 2 of them, keep the blade in line of perfect rotation on the blade housing insert attachment. Also in perfect symmetric activity is in order while pressure is applied to the blade during the cutting work done on any given materials.

11. The new tool on every model has where specified four 4, stability gears at degrees of 45, 135, 225, and 315 degree angles surrounding the blade. Of which are optional if the saw's model you buy comes with a blade stability circle.

Tropical Saws & Export Corporation ™

Perimeter Saw-Design Patent Application
Joe Nathan Brown-Inventor, Designer
Author, Engineering

THE **DESIGN PATENT**-APPLICATION, CLAIMS, AND
SPECIFICATION OF THE DESCRIPTIVE DRAWINGS

THE PERIMETER SAW © ®

12. THE DETAIL SUMMARY OF THE INVENTION - The Perimeter Saw model-A to Z gives rise to a large amount of possibilities when it comes to work, that will be done using it. The manufacturing of the Perimeter Saw will have stages, phases and steps that are per-determined for maximum output.

13. A. Stage 1 to 100-Tools used by hand; which are electric and non-electric to make the parts of the saw. Will be custom fitting and precision, and I will draw up some of them and add the needed tool to my process patent application when and where requested or required.

14. B. Phase 1 to 100-The equipment and machinery used to manufacture these 26 models of the saw will be made and built to the specifications of the parts made on them.

15. C. Steps 1 to 100-The buildings where these tools will be made. Will be safe for the workers, employees to accomplish the needed job and would be measured out, down to the floor planed square footage. So that taking steps of walking from one work station to the other to do a task, duty, job assignment, or taking a break for lunch and coffee, will be as easy as the steps 1, 2, 3 etc.

16. D. The new saw will need a prototype and working model made of each model, type and classification. So, as to add to the perspective of having used this product first, using the tests of its' working abilities are new leaving the page, and into a more weighty substance.

Tropical Saws & Export Corporation ™

Perimeter Saw-Design Patent Application
Joe Nathan Brown-Inventor, Designer
Author, Engineering

THE **DESIGN PATENT**-APPLICATION, CLAIMS AND
SPECIFICATIONS OF THE DESCRIPTIVE DRAWINGS

THE PERIMETER SAW © ®

17. THE CLAIMS NUMBER I-1. THE ORNAMENTAL DESIGNS AND FUCTION OF THE PERIMETER SAW.   The blades cutting edges, locations, lengths, sharpness, attaching angles, reductions and types make up claims number one. And pictures of the models A to Z will be attached either with those of model-X, or be filled alone as a separate patent claim. The views are the front, rear, top, bottom, left side and right side of the product.   The arrows on the pages indicate the part and piece being described.

18. The fully assembled saws will have all of the parts of the views present on it. And the parts; which are under or behind another part will be drawn up and displayed on a separate part of the page.   The blade rotating shaft has two designs one of them is made to turn the blade with gears on the circumference on the blade.   And the other is made to turn the blade from the front edge of it.

19. The electric motor installation placed in the center of the saw frame.   It will fit to specifications of the size of the opening for it.   And the horsepower to turn the blade of which it comprises also vary based on the size of the blade. Because the larger the blade the larger the motor would have to be.   Here are some of examples of the motor horsepower for the electric and gas powered motors: 1/8, 3/16, ¼, 5/8, ¾, 7/8, 1, 2, 3, 4, and 5 hp for the machinery and heavy equipment, heavy duty operational purposes.

20. The gas powered motor installations will also be placed in the saws' frame with mounting brackets and screw holes that fit to the specifications of the opening in the frame.   Its' rotation shaft will be customized to the blade it will turn.   The straight shaft connecting type and the elbow gear type, namely model-X, both have to be attached to a shaft made precise to the spacing of its' surrounding parts.

Tropical Saws & Export Corporation ™

Perimeter Saw-Design Patent Application
Joe Nathan Brown-Inventor, Designer
Author, Engineering

THE **DESIGN PATENT**-APPLICATION, CLAIMS AND
SPECIFICATIONS OF THE DESCRIPTIVE DRAWINGS

THE PERIMETER SAW © ®

THE CLAIMS NUMBER I-1, THE STRUCTURAL DESIGN

21. The blade's Gears for turning and how many of them, where they are located on the blade and the space between them make up the beginning of the claims filed for the design patent aspects noted herein after. Along with the structural design having a specific look: which changes per model A to Z, that. I Joe Nathan Brown have invented, a new tool called the Perimeter Saw.

22. The look of Design why each Model has a blade, holding apparatus and some means to turn the blade such as: the shaft, or motor drive. Since all of the models are specified to be between the lengths of 8" eight inches to 84" inches or seven feet. The part that adds the most to the lengths of the whole model saw are the handle and elongated frames, with extensions.

23. The Design Features because they make a difference, making it optional to buy or make a tool with at least two types of blade holding apparatus. Because the size of the blade's outer diameter will cause that a specific type of cover is made. It is important that it fits.

24. The dimensions and usages, for small and close spaces will allow that the saw sizes used for them be 8", 9", 10", 11", and 12" inches to make it between other materials around the materials to be cut. And a job that is a distance longer than 12" inches to reach an extension handle will be added to meet this demand.

25. The battery operated models will have an amperage storage capacity of 100 to 1000 amps consistently.

Tropical Saws & Export Corporation ™

Perimeter Saw-Design Patent Application
Joe Nathan Brown-Inventor, Designer, Author, Engineering

THE **DESIGN PATENT**-APPLICATION, CLAIMS AND
SPECIFICATIONS OF THE DESCRIPTIVE DRAWINGS

THE PERIMETER SAW © ®

26. It is possible to use the saw on pipes, tubing, rocks, nails, screws, wiring, roots of trees, branches, and all other substantial cutting needs. The tool that is the state of the art in design and utility is finally here. And it will make for cutting materials for sizing easier by it's' center blade rotations.

27. The equipment-as the sizes of the Perimeter Saws increases it will take on this term as equipment. And it will be used on larger jobs under more strenuous conditions. The Machinery-the heavy duty and industrial sizes of the perimeter saw, will be made to be mounted and attached to a table or on wheels. Because they are heavier they will be capable of completing jobs in the same class of usages.

28. The molding to make the 26 models of the perimeter saw will conform to all standard industrial markings of excellence. In that the sharpness of the blade, durability of the motor and all other parts will fit the criteria of engineering, professional standards.

29. The tools needed to manufacturer the Perimeter Saw, starting with the molding, sharpening, soldering, or welding using a soldering iron and grinding stones, are some activities; which will cause the new tool to be made. The screw driver, align wrench, pliers, and draft mans' precision cuts on the parts are a must to complete this project successfully.

Tropical Saws & Export Corporation ™

Perimeter Saw-Design Patent Application
Joe Nathan Brown-Inventor, Designer, Author, Engineering

THE **DESIGN PATENT**-APPLICATION, CLAIMS AND
SPECIFICATIONS OF THE DESCRIPTIVE DRAWINGS

THE PERIMETER SAW © ®

30. THE ABSTRACT OF THE DISCLOSURE DOCUMENT DEPOSIT REQUEST; which was filed on October 10, 2003, the first page of the documentation was in order. The date the first disclosure document was on October 10, 2003. In this form an introduction of my new "Perimeter Saw", that made its' first debut to the United States Patent and Trademarks Office.

31. While the saw would take more time to finish the design, utility, process patent and the manufacturing drawings. The basic blade housing the frame, that is its' structural content, have already been reported and disclosed, for documentation.

32. And the specifications, stages, phases, steps, order of assembly and order of operations are beginning to take shape through this design patent application. It covers the parts, pieces, and fragments of each connection and attachment known to be significant to obtaining an issuing of this patent.

33. The dimensions having first purposed, will in deed tell the perspective users, buyers, and sellers the needed information on this new product. With this informative claims and application the manufacturer and engineers can begin to size and spec. the drawings with perfect knowledge of the saws' presence. Therefore I am filing with confidence that the patent is the best way to introduce my new and innovative works of art and science, applied.

34. The original Design and Utility for my new product was model-X; which has the most types, styles and classifications in look, for a diversified portfolio on this new tool.

Tropical Saws & Export Corporation ™

Perimeter Saw-Design Patent Application
Joe Nathan Brown-Inventor, Designer, Author, Engineering

THE **DESIGN PATENT**-APPLICATION, CLAIMS AND
SPECIFICATIONS OF THE DESCRIPTIVE DRAWINGS

THE PERIMETER SAW © ®

THE ABSTRACT OF THE PROVISIONAL PATENT-PATENT PENDING OF: 4-14-2004

35. The date the first Provisional Patent was filed on April 14, 2004 and since then work has continued to finalize the drawings by myself, Joe Nathan Brown. With drafting, engineering, specifications and plans on how to make a prototype, working model and some unit product item for sale, when the patent processing is completed.

36. The Provisional Patent, was brought forth with the abbreviated drawings: which cover models- A, B, C, D, E, F, G, H, I, J, K, L, M, N, O, P, Q, R, S, T, U, V, W, X, Y, and Z. There by leading the way in the design patent to further show the evidence of developments on them; which has more solid lines, in shape and form.

37. The color differentials are: color each part, piece, and fragment. The color coding and assignments of the views being the exploded views, partial view, sectional views, alternate positional views, and modified forms views explain the 26 models of the Perimeter Saw. The actual design colors of the saws' parts are, brown, green, blue, yellow, orange, red, black, gold, silver, gray and multicolored expressions. It, the colors adds dimensions to the look of design and utility of the product and will have some means of conveying what the perimeter saw is. And as a part of my claims the colors are specified to identify my new product from all others on the market.

Tropical Saws & Export Corporation ™

Perimeter Saw-Design Patent Application
Joe Nathan Brown-Inventor, Designer, Author, Engineering

THE **DESIGN PATENT**-APPLICATION, CLAIMS AND
SPECIFICATIONS OF THE DESCRIPTIVE DRAWINGS

THE PERIMETER SAW © ®

38. THE DIMENSIONS ARE: as small as they can be made and as large as are feasible. Because with the depth and thickness of the blade, starting at 1/6"Th one sixteenth of an inch to the larger ones the depth being ½ one half an inch.

39. THE LENGTH ARE: are made to specifications on all models A to Z is measured from the top to the bottom of 8" inches to 84" inches or 7' seven feet.

40. THE WIDTH: of each model will start at 2 1/2" inches to 12" inches or one foot for the larger models of the saw.

41. THE DEPTH IN THICKNESS: are equal to the width, where specified depending on the components attachments and the frame 8" inches to 12" inches measurement.

42. THE HEIGHTS: of the 26 models of the Perimeter Saw, is equal to the lengths from the top of the saw to the bottom further most part.

43. THE WEIGHTS: of the Perimeter Saw will vary depending on the models sizing. And it will start at approximately 4 ounces to 150 pounds for the heavy industrial models.

44. THE COLORS: brown, black, red, yellow, orange, green, purple, white, and multicolored features can be added for definition.

Tropical Saws & Export Corporation ™

Perimeter **S**aw-**D**esign **P**atent **A**pplication
Joe Nathan Brown-Inventor, Designer, Author, Engineering

THE **DESIGN PATENT**-APPLICATION, CLAIMS AND
SPECIFICATIONS OF THE DESCRIPTIVE DRAWINGS

THE PERIMETER SAW © ®

THE NON-OBVIOUS FEATURES OF THE DESIGN ARE:

45. THE DURABILITY: of the Perimeter Saw is partly based on the materials needed to build them. It will be made of metal with water and dust proof sealing. They will be made of hard plastic unbreakable and water proof. They will be manufactured to the excess of basic standard for job performance and excellence. And all of the circuitry will be tightly assembled to prevent any damages to the users and applied to all forms of utility and ownership.

46. The blade has up to ten features of design and utility, 1-the cutting edges, 2-the cutting edges lengths, 3-the cutting edges sharpness, 4-the cutting edges angles to cut the materials, 5-the thickness of the blade, 6-the reduction and angle of them, 7-the durability, 8-the materials made to cover them and to make them, 9-the removal of only the cutting edges; which are spring or screw attached to the blade, 10-and the spacing between each cutting edges, that allows the wasted materials to fall through when the saw is in use.

47. The Blade Holding Apparatus, has ten features in the design and utility, it is a standard diameter being 1 ½", 1 ¾", 2", 2 1/4", 3", 3 1/4", 4", 41/4", 5", 5 1/4", 6", 6 1/4", 7", 7 1/4", 8" 8 1/4", 9" 9 1/4", 10", 10 1/4", 11", 11 1/4", 12", 12 1/4" and 12 1/2" inches and they will all increase in this and other specified lengths of diameter to cover the precise dimensions of a part on the front of the blade.

Perimeter Saw-Design Patent Application
Joe Nathan Brown-Inventor, Designer, Author, Engineering

THE **DESIGN PATENT**-APPLICATION, CLAIMS AND
SPECIFICATIONS OF THE DESCRIPTIVE DRAWINGS

THE PERIMETER SAW © ®

48. The horsepower of the motor and its' turning drive shaft is in sequence with the number of gears and their leverage activators; which cause the blade to cut. The spacing between the gears and the cutting edges separation will be equal, where specified. And because the gears will need the cutting edges in an even circle for best working results. And having them engineered this way will save on battery power per model used. And also when 120 volt electric cords are used the stress to the motor will be relieved in the same process of studies, testing, research and developments.

49. THE BLADES ARE THE MOST NON-OBVIOUS FEATURES OF THE PERIMETER SAW MAKING THEM PERFECT FOR PATENTING OF THIS NEW PRODUCT TECHNOLOGY SAW.

50. How the parts, pieces, and fragments of the invention are identified, numbered, lettered, and named.

51. A. Defining the parts-It was a job naming the parts of the saw, because it is new in its' features of the design. Starting with the cutting part that is called the blade and in all of the models the same name will be used to describe this part, and what it does.

52. B. Having said this, the blade has cutting edges, a circle round circumference, gears, and 1 or 2 blade stability circles on the rear of it. If the blade stability gears are used they will be positioned at 90, 180, 270, 360 or 45, 135, 225, and 315 degrees at points of the circumference that keep the blade from shifting and moving while it is making a perfect circular rotation.

Page 15. of 140.

Tropical Saws & Export Corporation ™

Perimeter Saw-Design Patent Application
Joe Nathan Brown-Inventor, Designer, Author, Engineering

THE **DESIGN PATENT**-APPLICATION, CLAIMS AND
SPECIFICATIONS OF THE DESCRIPTIVE DRAWINGS

THE PERIMETER SAW © ®

53. C. Defining the pieces-Is in this product nothing more than separating.

By making a part into smaller units of measurement, and lesser weights in

materials, they are reduced in size.

The pieces will be drawn to show all of the needed angles of them that will

matter the most to introduce a part.

54. Defining a fragment-Which are yet smaller units of the pieces such

As: each individual cutting edge of a possible one hundred 100. It tells how

Long they are the sharpness of them and the spacing in between them. The

Larger the blade the more cutting edges is added to it. The cutting edges a

part of the blade will take the angle in geometry that is best suited for the type

of materials it cuts. It does this by using the specific length. And since the

blade is In perfect symmetry the angle of the blades' Cutting edges will

correspond to the Degree of; which they are assigned. Such as: example the

angle of them may Start at 15 degrees to a feasible 55 Degrees, to match the

amount of materials that is to be removed from a Piece, of pipe or tubing it is

used on.

Tropical Saws & Export Corporation ™

Perimeter Saw © ®

Joe Nathan Brown-Inventor, Designer, Author, Engineering

DESIGN PATENT APPLICATION DRAWINGS
And Blue Prints

Page 17. of 140.

Tropical Saws & Export Corporation ™

Perimeter Saw-Design Patent Application
Joe Nathan Brown-Inventor, Designer, Author, Engineering
These are the CLAIMS AND SPECIFICATIONS.
February 23, 2009

1. FIG. I. This figure shown here is of the **Exploded View**, of the front of Model-X-1 of the fully assembled, Perimeter Saw, and pages 1 to 10. The parts; which are displayed and separated on this first page are:

The Order of Assembly: THE PERIMETER SAW Exploded View

1. Fig. I-a is the Component Insert Attachment & Frame.
2.Fig. I-b is the Blade Housing Insert Attachment, Top Frame, Friction Absorbing Pad.
3. Fig. I-c is the Blade Housing Insert Attachment, Top Frame, Friction Pad.
4. Fig. I-e This is the Blade Stability Gears and the Cylinder that holds it.
5. Fig. I-f This is the Blade Stability Gears and the Screws to mount it with.

1.1. The order of assembly of the Perimeter Saw begins on this page 1. It is within the drawings.

1.2. The Components Housing Insert Attachment & Frame 1. Fig I-a. Four holes are drilled into it for the mounting of the four Blade Stability Gears.

1.3. The Friction Absorbing Pads. 2. Fig. I-b and 3. Fig I-c is two possible installations. And Fig I-b as shown is mounted under the Blade Stability Gears 4 cylinders, but 3. Fig I-c is mounted on the elevation circle that allows only a 1/16 of an inch of this frame section to touch the blade, is a precise measurement for engineer purposes. On the larger model this sectional circle below the Friction Absorbing Pad will increase up to ½ an inch.

1.4. The Blade Stability Gears 1, 2, 3, & 4 are placed into the holes drilled for them by a cylinder shaft located on the base of it 4. Fig I-4d...

1.5. The Blade Stability Gears are tightened to the frame and Components Housing Insert Attachment, by four screws 4. Fig I-5e These Screws will have either align wrench heads or Phillips tip head for the use of two type screw tightening tools.

Tropical Saws & Export Corporation ™

Perimeter Saw © ®

Application Drawings and Prints
Joe Nathan Brown-Inventor, Designer, Artist, Author, Engineering
These are the CLAIMS & SPECIFICATIONS

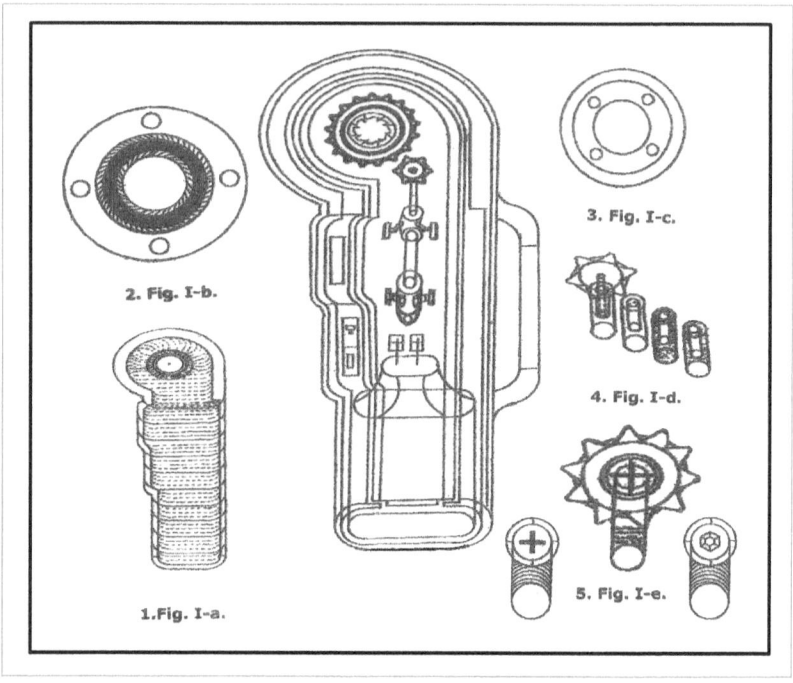

2. Fig. I-b.

3. Fig. I-c.

4. Fig. I-d.

5. Fig. I-e.

1.Fig. I-a.

1. FIG. I.

1. FIG. I This figure shown here is of the **Exploded View,** of the **Front of Model X-1** of the fully assembled, Perimeter Saw pages **18 to 41.** The parts are described in details on page **18** of the claims and specifications.

Page 19. of 140.

Tropical Saws & Export Corporation ™

Perimeter Saw-Design Patent Application
Joe Nathan Brown-Inventor, Designer, Author, Engineering

These are the CLAIMS AND SPECIFICATIONS I-1.
February 23, 2009

The Order of Assembly: THE PERIMETER SAW Exploded View

2. FIG. II. This figure shown here is of the **Exploded View**, of the **Front** of Model-X-1 of the fully assembled, Perimeter Saw pages 1. To 10.

1. Fig. II.-a The Electrical Wiring Connecting Terminals-Closed End.

2. Fig. II-b This is the Electrical Wiring Connecting Terminals-Open End.

3. Fig. II-c The 90 Degree Elbow Gear

4. Fig. II-d The Electric Motor DC & AC

2.1. The Electrical Wiring Connecting Terminals-Closed End is screw tightened to the Electric Motor and the Battery Plug in Terminals, Fig. II-a.

2.2. The Electrical Wiring Connecting Terminals-Open End is screw tightened to the Push Button Switch, Forward/Reverse Switch and the Position Locking Switch, Fig. II-b.

2.3. The 90 Degree Elbow Gear is Attached to the Motor by applied pressure in a firm gripping circle whole space, Fig. II-c.

2.4. The Electric Motor is installed after the Wiring Connection has been made to allow the most space to be used on these parts, Fig. II-d.

Page 20. of 140.

Tropical Saws & Export Corporation ™

Perimeter Saw © ®

Application Drawings and Prints
Joe Nathan Brown-Inventor, Designer, Artist, Author, Engineering
These are the CLAIMS & SPECIFICATIONS

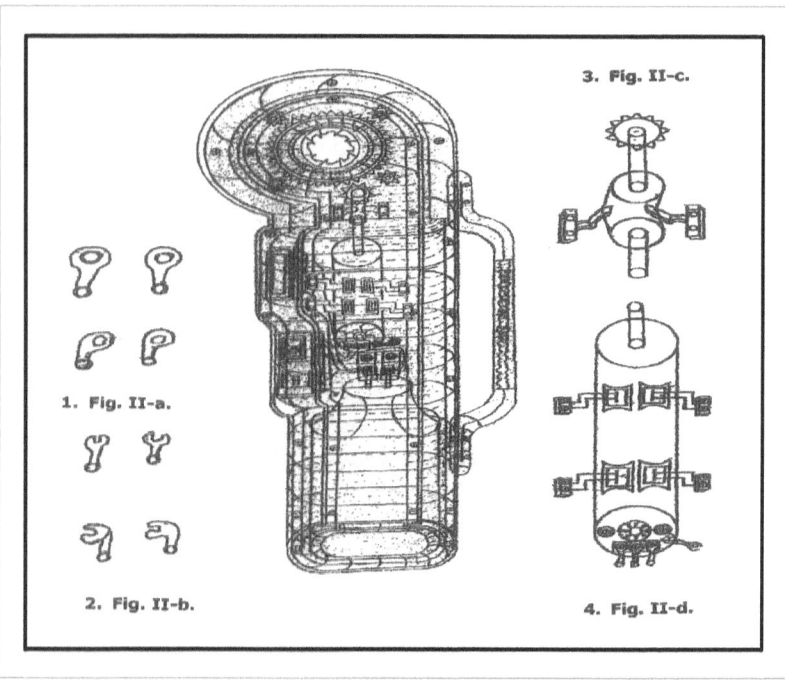

3. Fig. II-c.

1. Fig. II-a.

2. Fig. II-b.

4. Fig. II-d.

2. FIG. II.

2. FIG. II This figure shown here is of the **Exploded View**, of the Front of **Model X-1** of the fully assembled, **Perimeter Saw** pages **18 to 41.** The parts are described in details on page **20** of the claims and specifications.

Tropical Saws & Export Corporation ™

Perimeter **S**aw-**D**esign **P**atent **A**pplication
Joe Nathan Brown-Inventor, Designer, Author, Engineering
These are the CLAIMS AND SPECIFICATIONS I-1.
February 23, 2009

The Order of Assembly: THE PERIMETER SAW Exploded View

3. FIG. III. This figure shown here is of the **Exploded View**, of the **Front** of Model-X-1 the fully assembled, Perimeter Saw. It shows the fully assembled saw with 30 to 60 screws attaching the parts together.

1. Fig. III-a is the Electric Components Connecting Wiring.

2. Fig. III-b this is The Battery Connection Plug-in Terminals & 120 volt Electric Cord Splicing Terminals.

3. Fig. III-c The Rubber Insulation below the Battery Connection Plug in Terminals.

4. Fig. III-d The Thermal Plastic Insulation over the Battery Connection Plug In Terminals.

3.1. The Electric Component Connecting Wiring is attached to the Battery Plug in Terminals with the Electrical Wiring Connecting Terminals-Close & Open End Terminals, Fig. III-a.

3.2. The Battery Connection Plug in Terminals can be attached to the Component Housing Insert Attachment, by two screws, Fig. III-b.

3.3. The Rubber Insulation below the Battery Connection Plug in Terminals is place just below the Battery Plug in Terminals to stop the grounding of electric part near by, Fig. III-c.

3.4. The Thermal Plastic Insulation over the Battery Connection Plug in Terminal is installed before the Plug- in Terminals is attached to the Frame, Fig. III-d.

Tropical Saws & Export Corporation ™

Perimeter Saw © ®

Application Drawings and Prints
Joe Nathan Brown-Inventor, Designer, Artist, Author, Engineering
These are the CLAIMS & SPECIFICATIONS

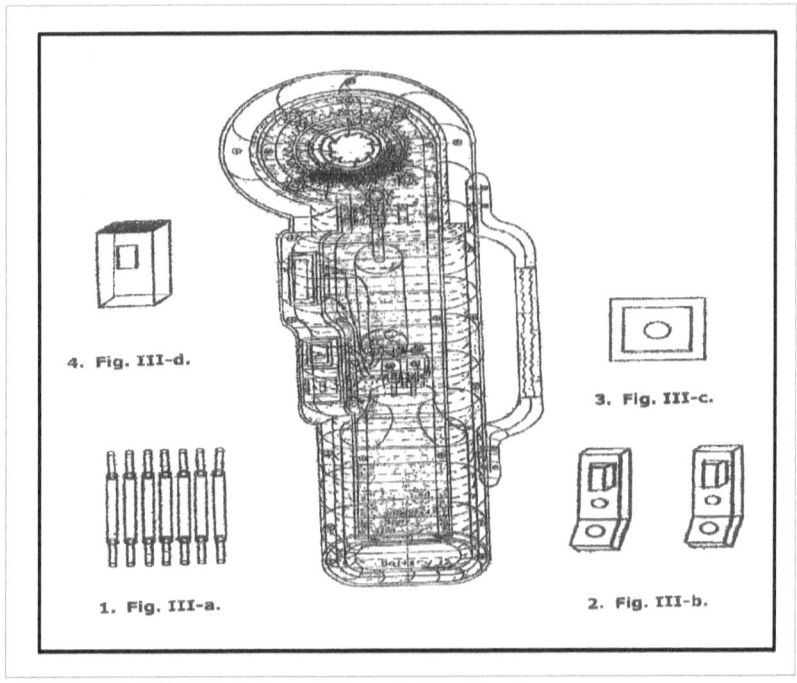

4. Fig. III-d.

3. Fig. III-c.

1. Fig. III-a.

2. Fig. III-b.

3. FIG. III.

3. FIG. III This figure shown here is of the **Exploded View,** of the **Front of Model** X-1 of the fully assembled, **Perimeter Saw** pages **18 to 41.** The parts are described in details on page **22** of the claims and specifications.

Page 23. of 140.

Tropical Saws & Export Corporation ™

Perimeter Saw-Design Patent Application
Joe Nathan Brown-Inventor, Designer, Author, Engineering

These are the CLAIMS AND SPECIFICATIONS I-1.
May 27, 2009

The Order of Assembly: THE PERIMETER SAW Exploded View

4. FIG. IV. This figure shown here is of the **Exploded View**, the **Front** of Model-X-1 of the fully assembled, and Perimeter Saw.

1. Fig. IV.-a This is the Component Housing Insert Attachment Front, with the opening for the Components Center Cover.

2. Fig. IV.-b Is the Push Button Electrical Switch low and High Voltage.

3. Fig. IV-C1 This is the Forward and Reverse Switch, upper location.

4. Fig. IV-C2 This is the Position Locking on/off Switch, lower location.

4.1. The Component Housing Insert Attachment, Front with the opening for the Components Center Cover, is installed to the rear Component Housing Insert Attachment, by 10 to 13 screws, Fig. IV-a.

4.2. The Push Button Electrical Switch Low and High Voltage are installed in the frame, with two screws, Fig. IV-b.

4.3. The Forward and Reverse Switch, upper location being installed by two screws since this is a combination device, Fig. IV-c.

4.4. The Position locking on/off Switch, Lower location a part of the combination device is installed by two screws, Fig. I-d.

Page 24. of 140.

Tropical Saws & Export Corporation ™

Perimeter Saw © ®

Application Drawings and Prints
Joe Nathan Brown-Inventor, Designer, Artist, Author, Engineering
These are the CLAIMS & SPECIFICATIONS

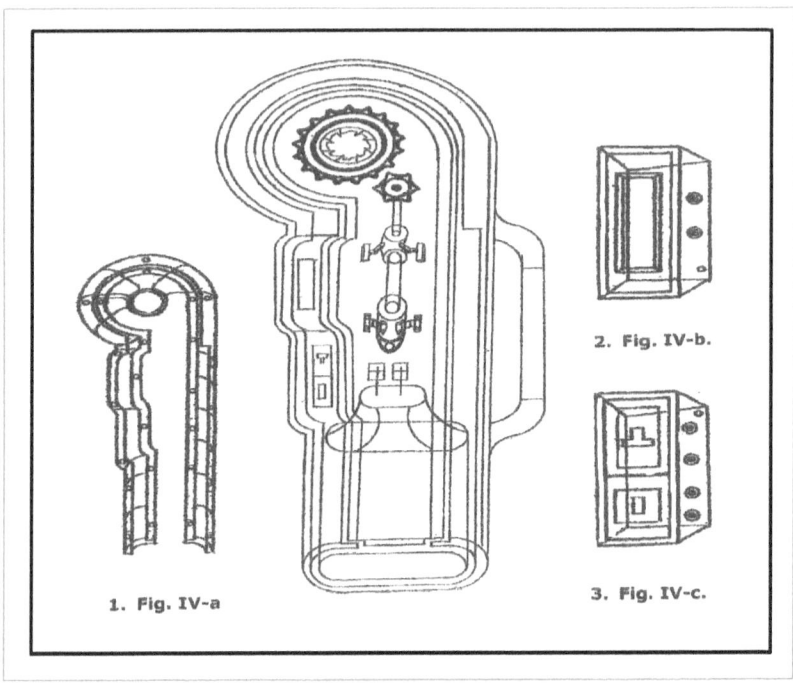

1. Fig. IV-a

2. Fig. IV-b.

3. Fig. IV-c.

4. FIG. IV.

4. FIG. IV This figure shown here is of the **Exploded View,** of the **Front of Model** X-1 of the fully assembled, **Perimeter Saw** pages **18 to 41**. The parts are described in details on page **24** of the claims and specifications.

Page 25. of 140.

Tropical Saws & Export Corporation ™

Perimeter Saw-Design Patent Application
Joe Nathan Brown-Inventor, Designer, Author, Engineering

These are the CLAIMS AND SPECIFICATIONS I-1.
May 27, 2009

The Order of Assembly: THE PERIMETER SAW, Exploded View

5. FIG. V. This figure shown here is of the **Exploded View**, of the **Front** of Model-X-1 of the fully assembled, Perimeter Saw.   The parts are the Components Housing Insert Attachment, with screw holes. The Components Housing Insert Attachment is a three part mounting means and the frame.

1. Fig. V-a is the piece of the Bottom Components Cover, section on the Frame, for 4 screws.

2. Fig. V-b The Bottom Components Cover, with four screws attachment holes.

3. Fig. V-c The 12v, 14v, 18v, 24v & 48 Volt DC current Storage Battery are used in the saw.

5.1. The Bottom Component Cover, a piece that is a part of the two adjoining parts the upper and lower Components Housing Insert Attachments, made with 2 to 4 threaded screw holes.

5.2. The Bottom Component Cover, with 4 screw holes is mounted to the frame after the battery has been installed, Fig. V-b.

5.3. The 12volt, 14volt, 18volt, 24volt and the 48 volt DC current storage Battery, is installed in the space provided for them as they will increase in size and storage capacity as the Perimeter Saw does. They will be made for the 12 different sizes of Saw types, Styles and Classifications.

Tropical Saws & Export Corporation ™

Perimeter Saw © ®

Application Drawings and Prints
Joe Nathan Brown-Inventor, Designer, Artist, Author, Engineering
These are the CLAIMS & SPECIFICATIONS

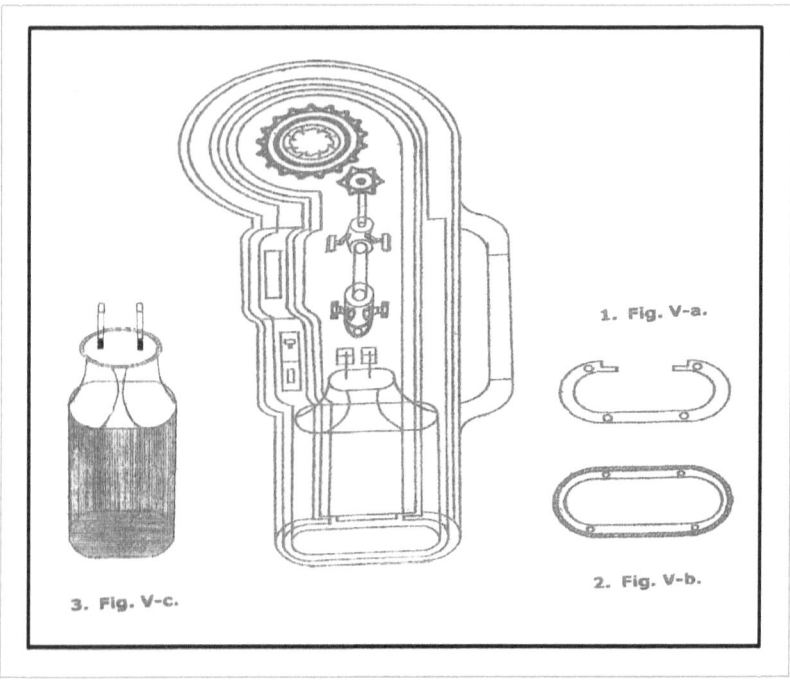

1. Fig. V-a.

2. Fig. V-b.

3. Fig. V-c.

5. FIG. V.

5. FIG. V This figure shown here is of the **Exploded View,** of the Front of **Model** X-1 of the fully assembled, **Perimeter Saw** pages **18 to 41.** The parts are described in details on page **26** of the claims and specifications.

Page 27. of 140.

Tropical Saws & Export Corporation ™

Perimeter **S**aw-**D**esign **P**atent **A**pplication
Joe Nathan Brown-Inventor, Designer, Author, Engineering

These are the CLAIMS AND SPECIFICATIONS I-1.
February 24, 2009

The Order of Assembly: THE PERIMETER SAW, Exploded View

6. FIG. VI. This figure shown here is of the **Exploded View**, of the **Front** of Model-X-1 of the fully assembled, Perimeter Saw. The part in this view is the Utility Handle.

1. Fig. VI-a The Utility & Storage Handle will have the sizing to fit the hand and fingers. It will be made in small, medium and large with a connection spacing of 1″, 2″, 3″, 4″, 5″, 6″, 7″, 8″, 9″, and 10″, inches between the gripping Pad in the middle of it. This opening for the hand extends to the screw attaching holes in the frame. It is attached and mounted to the section of the saw where the front and rear Components Housing Insert Attachment meet and are adjoined on the right side. It also is counter sunk into the two sections for a recessed finish, Fig. VI-a.   The handle will have griping rubber and fibers that help the holding capacity of the user.The handle is made of materials that are strong enough to resist the pressures of holding it while cutting rugged materials.

Tropical Saws & Export Corporation ™

Perimeter Saw © ®

Application Drawings and Prints
Joe Nathan Brown-Inventor, Designer, Artist, Author, Engineering
These are the CLAIMS & SPECIFICATIONS

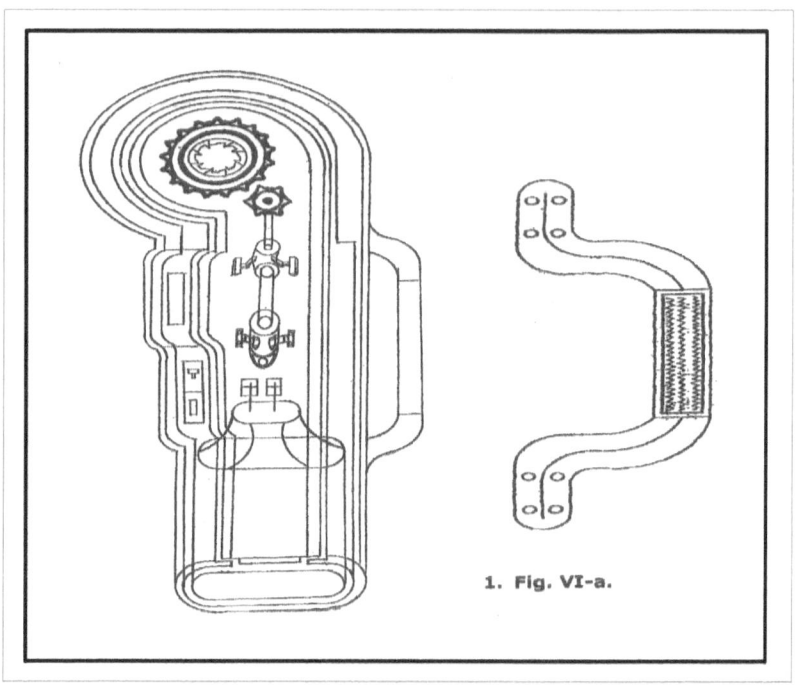

1. Fig. VI-a.

6. FIG. VI.

6. FIG. VI This figure shown here is of the **Exploded View,** of the Front of Model X-1 of the fully assembled, **Perimeter Saw** pages **18 to 41.** The parts are described in details on page **28** of the claims and specifications.

Page 29. of 140.

Tropical Saws & Export Corporation ™

Perimeter Saw-Design Patent Application
Joe Nathan Brown-Inventor, Designer, Artist, Author, Engineering

These are the CLAIMS AND SPECIFICATIONS I-1.

7. FIG. VII-1. This figure shown here is of the **Exploded View**, of the **Front** of Model-X-1 of the fully assembled, Perimeter Saw.

7.1.	Arrow # 1.	The Blade Stability Gear
7.2.	Arrow # 2	Blade"s Cutting Edges, Stability Circle & Friction Pad
7.3.	Arrow # 3.	The Rear Components Insert Attachment & Frame
7.4.	Arrow # 4.	The Blade Stability Gear
7.5.	Arrow # 5.	The Push Button on/off Switch
7.6.	Arrow # 6.	The Forward & Reverse Switch
7.7	Arrow # 7.	The Position Locking on/off Switch
7.8.	Arrow # 8.	The Electric Components Connecting Wiring
7.9.	Arrow # 9.	The DC Battery 12v, 14v, 24v 48volts
7.10.	Arrow #10.	The Blade Stability Gear
7.11.	Arrow #11.	The Front Components Insert Attachment
7.12.	Arrow #12.	The Blade Stability Gear
7.13.	Arrow #13.	The 90 Degree Elbow Gear
7.14.	Arrow #14.	The Utility Handle
7.15.	Arrow #15.	The Electric, Gas, or Air Compressor Motor
7.16.	Arrow #16.	This is the Battery Connecting Plug- in Terminals
7.17.	Arrow #17.	The Components Insert Attachment, Center Cover
7.18.	Arrow #18.	The Components Mounting Screws
7.19.	Arrow #19.	The Bottom Components Cover

Tropical Saws & Export Corporation ™

Perimeter Saw © ®

Application Drawings and Prints
Joe Nathan Brown-Inventor, Designer, Artist, Author, Engineering
These are the CLAIMS & SPECIFICATIONS

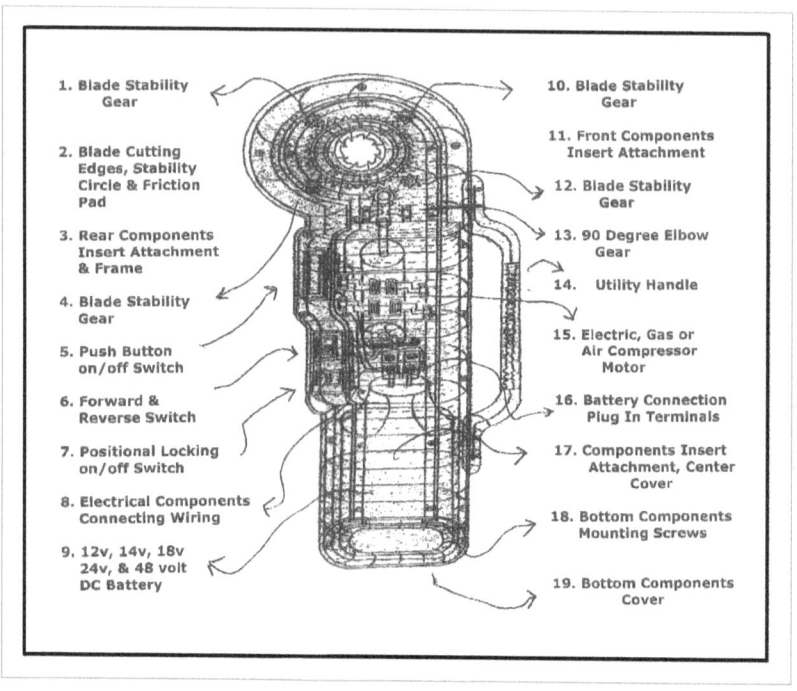

1. Blade Stability Gear

2. Blade Cutting Edges, Stability Circle & Friction Pad

3. Rear Components Insert Attachment & Frame

4. Blade Stability Gear

5. Push Button on/off Switch

6. Forward & Reverse Switch

7. Positional Locking on/off Switch

8. Electrical Components Connecting Wiring

9. 12v, 14v, 18v 24v, & 48 volt DC Battery

10. Blade Stability Gear

11. Front Components Insert Attachment

12. Blade Stability Gear

13. 90 Degree Elbow Gear

14. Utility Handle

15. Electric, Gas or Air Compressor Motor

16. Battery Connection Plug In Terminals

17. Components Insert Attachment, Center Cover

18. Bottom Components Mounting Screws

19. Bottom Components Cover

7. FIG. VII-1.

7. FIG. VII-1 This figure shown here is of the **Exploded View,** of the **Front of Model X-1** of the fully assembled, **Perimeter Saw** pages **18 to 41.** The parts are described in details on page **30** of the claims and specifications.

Tropical Saws & Export Corporation ™

Perimeter Saw-Design Patent Application
Joe Nathan Brown-Inventor, Designer, Author, Engineering

These are the CLAIMS AND SPECIFICATIONS I-1.

7. FIG. VII-2 This figure shown here is of the **Exploded View**, of the **Front** of Model-X-1 of the fully assembled, Perimeter Saw.

7.1. Arrow # 1.	The Blade Stability Gear
7.2. Arrow # 2	Blade"s Cutting Edges, Stability Circle & Friction Pad
7.3. Arrow # 3.	The Rear Components Insert Attachment & Frame
7.4. Arrow # 4.	The Blade Stability Gear
7.5. Arrow # 5.	The Push Button on/off Switch
7.6. Arrow # 6.	The Forward & Reverse Switch
7.7 Arrow # 7.	The Position Locking on/off Switch
7.8. Arrow # 8.	The Electric Components Connecting Wiring
7.9. Arrow # 9.	The DC Battery 12v, 14v, 24v 48volts
7.10. Arrow #10.	The Blade Stability Gear
7.11. Arrow #11.	The Front Components Insert Attachment
7.12. Arrow #12.	The Blade Stability Gear
7.13. Arrow #13.	The 90 Degree Elbow Gear
7.14. Arrow #14.	The Utility Handle
7.15. Arrow #15.	The Electric, Gas, or Air Compressor Motor
7.16. Arrow #16.	This is the Battery Connecting Plug- in Terminals
7.17. Arrow #17.	The Components Insert Attachment, Center Cover
7.18. Arrow #18.	The Components Mounting Screws
7.19. Arrow #19.	The Bottom Components Cover

Tropical Saws & Export Corporation ™

Perimeter Saw © ®

Application Drawings and Prints
Joe Nathan Brown-Inventor, Designer, Artist, Author, Engineering
These are the CLAIMS & SPECIFICATIONS

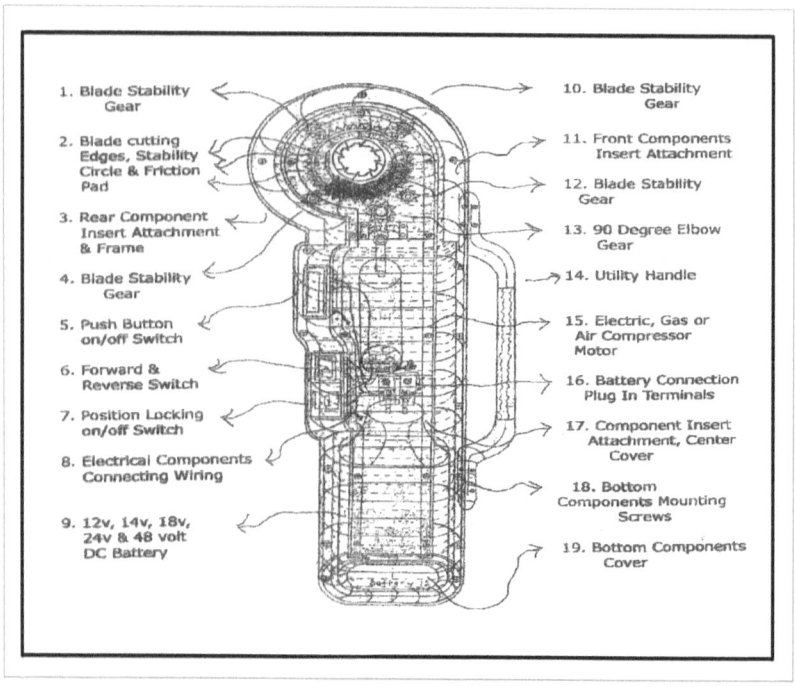

1. Blade Stability Gear
2. Blade cutting Edges, Stability Circle & Friction Pad
3. Rear Component Insert Attachment & Frame
4. Blade Stability Gear
5. Push Button on/off Switch
6. Forward & Reverse Switch
7. Position Locking on/off Switch
8. Electrical Components Connecting Wiring
9. 12v, 14v, 18v, 24v & 48 volt DC Battery

10. Blade Stability Gear
11. Front Components Insert Attachment
12. Blade Stability Gear
13. 90 Degree Elbow Gear
14. Utility Handle
15. Electric, Gas or Air Compressor Motor
16. Battery Connection Plug In Terminals
17. Component Insert Attachment, Center Cover
18. Bottom Components Mounting Screws
19. Bottom Components Cover

7. FIG. VIII-2.

7. FIG. VIII-2 This figure shown here is of the **Exploded View,** of the **Front of Model X-1** of the fully assembled, **Perimeter Saw** pages **18 to 41**. The parts are described in details on page **32** of the claims and specifications.

Page 33. of 140.

Tropical Saws & Export Corporation ™

Perimeter Saw-Design Patent Application
Joe Nathan Brown-Inventor, Designer, Author, Engineering

These are the CLAIMS AND SPECIFICATIONS I-1.

7. FIG. VII-3 This figure shown here is of the **Exploded View**, of the **Front** of Model-X-1 of the fully assembled, Perimeter Saw.

7.1. Arrow # 1.	The Blade Stability Gear
7.2. Arrow # 2	Blade"s Cutting Edges, Stability Circle & Friction Pad
7.3. Arrow # 3.	The Rear Components Insert Attachment & Frame
7.4. Arrow # 4.	The Blade Stability Gear
7.5. Arrow # 5.	The Push Button on/off Switch
7.6. Arrow # 6.	The Forward & Reverse Switch
7.7 Arrow # 7.	The Position Locking on/off Switch
7.8. Arrow # 8.	The Electric Components Connecting Wiring
7.9. Arrow # 9.	The DC Battery 12v, 14v, 24v 48volts
7.10. Arrow #10.	The Blade Stability Gear
7.11. Arrow #11.	The Front Components Insert Attachment
7.12. Arrow #12.	The Blade Stability Gear
7.13. Arrow #13.	The 90 Degree Elbow Gear
7.14. Arrow #14.	The Utility Handle
7.15. Arrow #15.	The Electric, Gas, or Air Compressor Motor
7.16. Arrow #16.	This is the Battery Connecting Plug- in Terminals
7.17. Arrow #17.	The Components Insert Attachment, Center Cover
7.18. Arrow #18.	The Components Mounting Screws
7.19. Arrow #19.	The Bottom Components Cover

Tropical Saws & Export Corporation ™

Perimeter Saw © ®

Application Drawings and Prints
Joe Nathan Brown-Inventor, Designer, Artist, Author, Engineering
These are the CLAIMS & SPECIFICATIONS

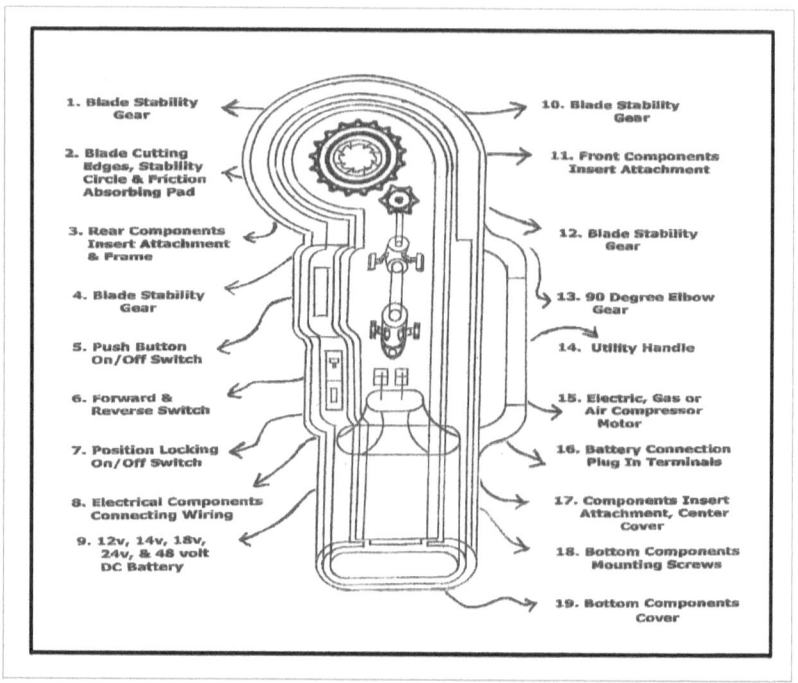

1. Blade Stability Gear
2. Blade Cutting Edges, Stability Circle & Friction Absorbing Pad
3. Rear Components Insert Attachment & Frame
4. Blade Stability Gear
5. Push Button On/Off Switch
6. Forward & Reverse Switch
7. Position Locking On/Off Switch
8. Electrical Components Connecting Wiring
9. 12v, 14v, 18v, 24v, & 48 volt DC Battery

10. Blade Stability Gear
11. Front Components Insert Attachment
12. Blade Stability Gear
13. 90 Degree Elbow Gear
14. Utility Handle
15. Electric, Gas or Air Compressor Motor
16. Battery Connection Plug In Terminals
17. Components Insert Attachment, Center Cover
18. Bottom Components Mounting Screws
19. Bottom Components Cover

7. FIG. VII-3.

7. FIG. VII-3 This figure shown here is of the **Exploded View,** of the **Front of Model X-1** of the fully assembled, **Perimeter Saw** pages **18 to 41.** The parts are described in details on page **34** of the claims and specifications.

Page 35. of 140.

Tropical Saws & Export Corporation ™

Perimeter **S**aw-**D**esign **P**atent **A**pplication
Joe Nathan Brown-Inventor, Designer, Author, Engineering

These are the CLAIMS AND SPECIFICATIONS I-1.
May 27, 2009

The Order of Assembly: THE PERIMETER SAW Exploded View

8. FIG. VIII. This figure shown here is of the **Exploded View**, of the **Front** of Model-X-1 of the fully assembled, Perimeter Saw. The parts in this view are the Switches, Push Button, Forward/Reverse and Locking Positional type.   This figure shown here adds to the Claim number 1 with the saw in the 120 Volt Electric Cords for continuous operations and use.

1. Fig. VIII-A This is the Claim number I-1, the 120 Volt Continuous Duty Operating Model-X-1, shown on page 37.

8.1. The Claim number I-1 is of the 120 volt model of the Perimeter Saw;

that will be made for light and heavy duty usages. It will have a 6 and 8

Feet cord attached to the bottom, in a hole of the Component Bottom Cover.

This model-X, will be made to specification such as the Push Button Switch

Having the capacity to cause the motor to go high or low, described here as

The variable speed switch is one of the components. All of the standard sizes

of the blade apply here. The dimensions the blade has will cause the frame to

be made from small to large and they are: Blade sizes 4″, 8″, 12″, 16″, 20″, 24″,

28″, 32″, 36″, 40″, 44″, and 48″ Inches in the blades outer diameter, are a

standard size.

Tropical Saws & Export Corporation ™

Perimeter Saw © ®

Application Drawings and Prints
Joe Nathan Brown-Inventor, Designer, Artist, Author, Engineering
These are the CLAIMS & SPECIFICATIONS

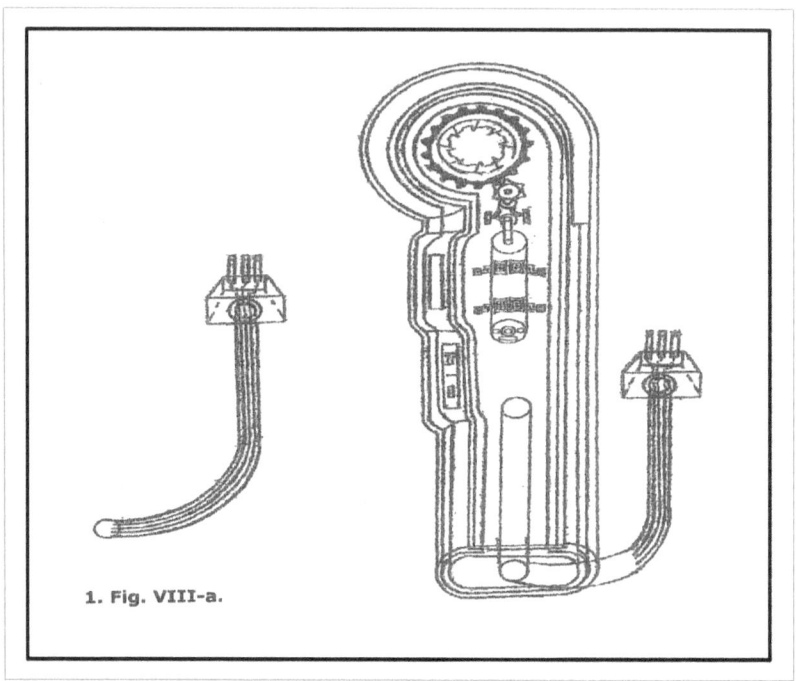

1. Fig. VIII-a.

8. FIG. VIII.

8. FIG. VIII This figure shown here is of the **Exploded View,** of the **Front of Model X-1** of the fully assembled, **Perimeter Saw** pages **18 to 41**. The parts are described in details on page **36** of the claims and specifications.

Page 37. of 140.

Tropical Saws & Export Corporation ™

Perimeter Saw-Design Patent Application
Joe Nathan Brown-Inventor, Designer, Author, Engineering

These are the CLAIMS AND SPECIFICATIONS I-1.
February 23, 2009

The Order of Assembly: THE PERIMETER SAW, Exploded View

9. FIG. IX. This figure shown here is of the **Exploded View**, of the **Front** of Model-X-1 of the fully assembled, Perimeter Saw. This figure shows how the bottom of the saw looks displaying the spacing between the Front, Component Housing Insert Attachment and the Rear Components Housing Insert Attachment or the Frame.

1. Fig.-IX-a. The Perimeter Saw Reduced to show the ability to make it into a precision tool.

2. Fig-IX-b. The Push Button Switch, Opening is shown.

3. Fig-IX-c. The Forward & Reverse Switch and Position Locking Switch, Opening are controlling switches.

9.1. The Perimeter Saw is reduced here in size and the order of assembly here represents the ability to make the product smaller. This process can be done in a larger model unit and then applied to this drawing, Fig. IX-a.

9.2. The Push Button Switch, Opening is created by the adjoining of the Upper and Lower Components Insert Attachment and will have the appearance as shown here, Fig. IX-b.

9.3. The Forward/Reverse Switch and Position Locking Switch, Opening is formed by the upper and Lower Components Insert Attachment and will have the appearance as shown here, Fig. IX-c.

Page 38. of 140.

Tropical Saws & Export Corporation ™

Perimeter Saw © ®

Application Drawings and Prints
Joe Nathan Brown-Inventor, Designer, Artist, Author, Engineering
These are the CLAIMS & SPECIFICATIONS

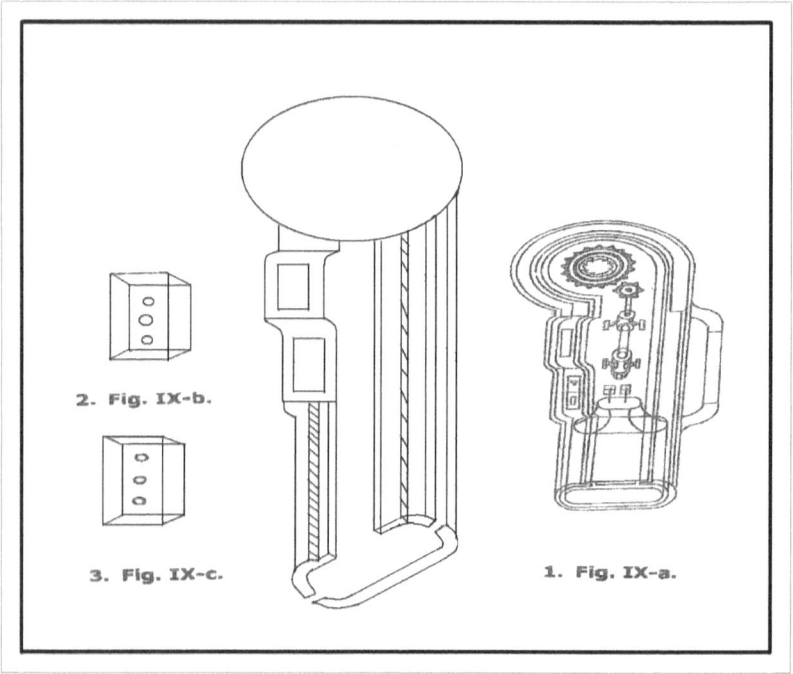

2. Fig. IX-b.

3. Fig. IX-c.

1. Fig. IX-a.

9. FIG. IX.

9. FIG. IX This figure shown here is of the **Exploded View,** of the **Front of Model X-1** of the fully assembled, **Perimeter Saw** pages **18 to 41.** The parts are described in details on page **38** of the claims and specifications.

Page 39. of 140.

Tropical Saws & Export Corporation ™

Perimeter **S**aw-**D**esign **P**atent **A**pplication
Joe Nathan Brown-Inventor, Designer, Author, Engineering

These are the CLAIMS AND SPECIFICATIONS I-1.
February 23, 2009

The Order of Assembly: THE PERIMETER SAW, Exploded View

10. FIG. X. This figure shown here is of the **Exploded View**, of the **Front** of Model-X-1 of the fully assembled, Perimeter Saw. The view of the Top region has the Blade Stability Gears and the 90 Degree Elbow Gear.

The 90 Degree Elbow Gear is installed to the motor it will be attached to the Components Housing Insert Attachment by 4 screws.

The Blade Stability Gears are a total of 4, they will be set and installed at symmetric location evenly spaced surrounding the blade to keep the blade moving in perfect rotation, when pressure is applied to them by the applied force of the material being cut.

1. Fig. X.-a The Fully assembled Perimeter Saw, reduced in Size.   The Perimeter Saw is reduced here in size and the order of assembly here represents the ability to make the product smaller. This process can be done in a larger model unit and the applied to this drawing, Fig. IX-a.

2. Fig. X-b The Fully assembled Perimeter Saw, reduced in Size.   The Perimeter Saw is reduced here in size and the order of assembly here represents the ability to make the product smaller. This process can be done in a larger model unit and the applied to this drawing, Fig. IX-a.

Tropical Saws & Export Corporation ™

Perimeter Saw © ®

Application Drawings and Prints
Joe Nathan Brown-Inventor, Designer, Artist, Author, Engineering
These are the CLAIMS & SPECIFICATIONS

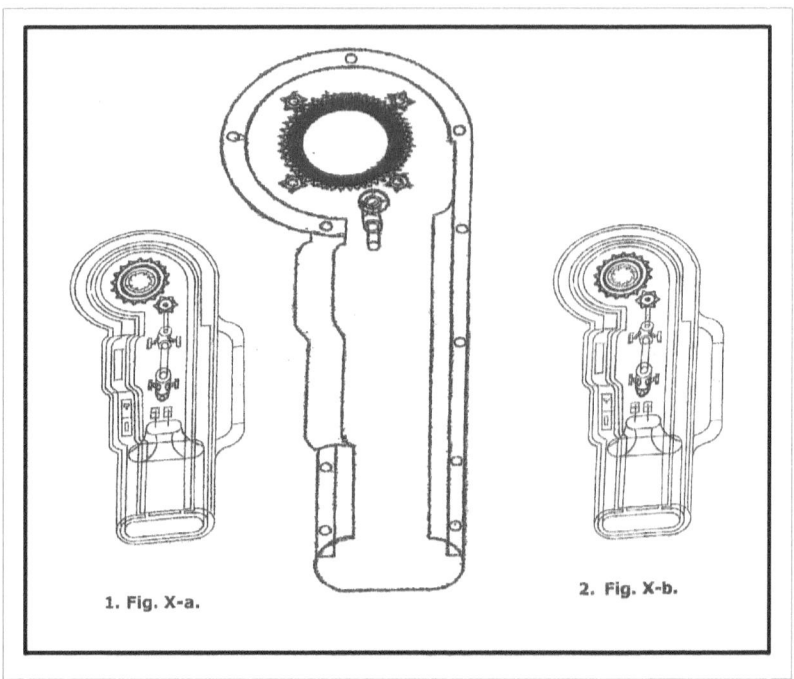

1. Fig. X-a.

2. Fig. X-b.

10. FIG. X.

10. FIG. X This figure shown here is of the **Exploded View,** of the **Front of Model X-1** of the fully assembled, **Perimeter Saw** pages **18 to 41.** The parts are described in details on **p**age **40** of the claims and specifications.

Page 41. of 140.

Tropical Saws & Export Corporation ™

Perimeter Saw-Design Patent Application
Joe Nathan Brown-Inventor, Designer, Author, Engineering

These are the CLAIMS AND SPECIFICATIONS I-1.
February 23, 2009

Parts Removed to Make: Partial Views

11. FIG. XI   This figure shown here is of the **Partial View**, of the **Front** Model-X-1 of the fully assembled, Perimeter Saw, and it's Components Housing Insert Attachment & Frame in the center of the page. This is an open look of the saw without any components attached. The upper region shows the Blade Housing Insert Attachment, molded into this top section. The raised platform is circular shaped and it will be covered by the Components Housing Insert Attachment, Friction Absorbing Pad.

1. Fig. XI.-a   The Fully Assembled Perimeter Saw, Reduced in Size, to show a smaller precise saw. The Perimeter Saw is reduced here in size and the order of assembly here represents the ability to make the product smaller.   This process can be done in a larger model unit and the applied to this drawing, Fig. IX-a

2. Fig. XI-b The Fully Assembled Perimeter Saw, Reduced in Size, to show a smaller precise saw. The Perimeter Saw is reduced here in size and the order of assembly here represents the ability to make the product smaller.   This process can be done in a larger model unit and the applied to this drawing, Fig. IX-a

Tropical Saws & Export Corporation ™

Perimeter Saw © ®

Application Drawings and Prints
Joe Nathan Brown-Inventor, Designer, Artist, Author, Engineering
These are the CLAIMS & SPECIFICATIONS

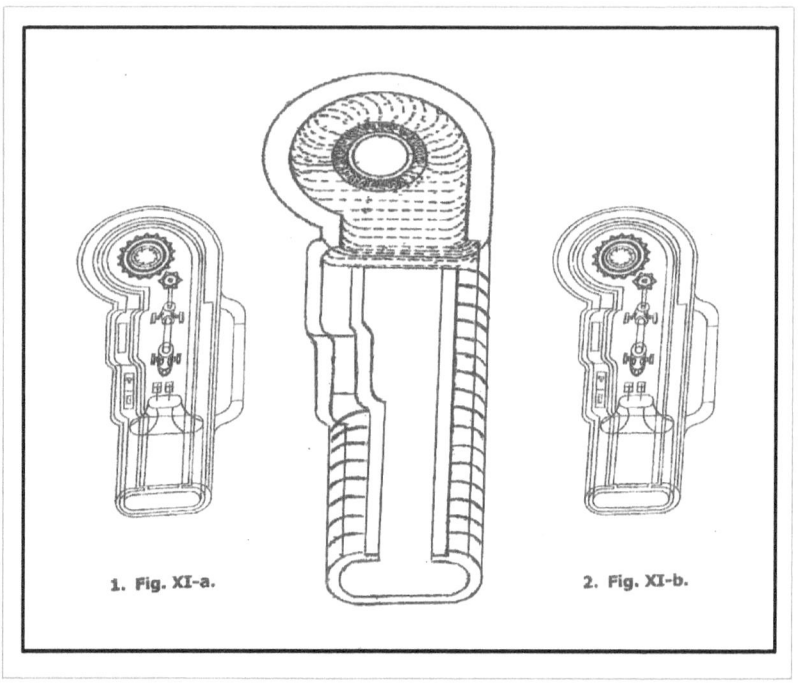

1. Fig. XI-a. 2. Fig. XI-b.

11. FIG. XI.

11. FIG. XI This figure shown here is of the **Partial View,** of the **Front of Model X-1** of the fully assembled, **Perimeter Saw** pages **42 to 51.** The parts are described in details on page **42** of the claims and specifications.

Page 43. of 140.

Tropical Saws & Export Corporation ™

Perimeter Saw-Design Patent Application
Joe Nathan Brown-Inventor, Designer, Author, Engineering

These are the CLAIMS AND SPECIFICATIONS I-1.
February 23, 2009

Parts Removed to Make: Partial Views

12. FIG. XII This figure shown here, is of the **Partial View**, of the **Front** of Model-X-1 of the fully assembled, Perimeter Saw and its' Component Housing Insert Attachment, in the center of the page. The part shown here if made as a separate piece, without the bottom piece, could be used to fit in between the Component Housing Insert attachment, front and rear section. It would have 11 screw threaded holes on the left and right sides and 4 screw threaded holes on the Component Bottom Cover space. The view shows the top blade section being perfectly round and circular in the region where the blade will be installed. It shows the space between the top blade section and the location of the first Push Button Switch. This space will be made large enough to add length into the entire Perimeter Saws' Frame. On the Model-X-1a it will be a length between 1/2" to 1" inch for the blades capacity of in diameter of 4" inches. Here are the specification for the space on the left side of the frame between the blade section and the Push Button Switch.

Blade's Diameter, Left Frame Space Space		Blade's Diameter, Left Frame Space	
1. 4"	1/2", 3/4", 1"	7. 28"	4", 4 1/2", 4 3/4"
2. 8"	3/4", 1", 1 1/2"	8. 32"	5", 5 1/2", 5 3/4"
3. 12"	1", 1 1/2", 1 3/4"	9. 36"	6", 6 1/2", 6 3/4"
4. 16"	1 3/4", 2", 2 1/2"	10. 40"	7", 7 1/2", 7 3/4"
5. 20"	2", 2 1/2", 2 3/4"	11. 44"	7", 7 1/2", 7 3/4"
6. 24"	3", 3/12", 3 3/4"	12. 48"	7", 7 1/2", 7 3/4"

1. **Fig. XII.-a, 2. Fig. XII-b** The Fully Assembled Perimeter Saw, Reduced in Size. This process can be done in larger model units and then applied to this drawing.

Page 44. of 140.

Tropical Saws & Export Corporation ™

Perimeter Saw © ®

Application Drawings and Prints
Joe Nathan Brown-Inventor, Designer, Author, Engineering

These are the CLAIMS & SPEICIFICATIONS I-1.

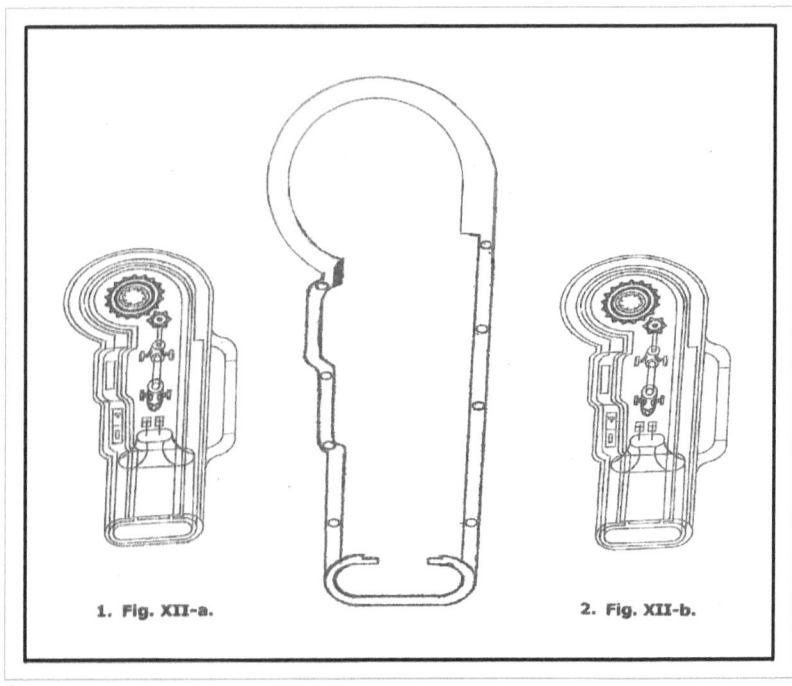

1. Fig. XII-a.   2. Fig. XII-b.

12. FIG. XII.

12. FIG. XII. This figure shown here is of the, **Partial View** of the **Front of Model X-1** of the fully assembled **Perimeter Saw,** pages **42** to **51**. The parts are described in detail on **page 44** of the Claims and Specifications.

Page 45. of 140.

Tropical Saws & Export Corporation ™

Perimeter Saw-Design Patent Application
Joe Nathan Brown-Inventor, Designer, Author, Engineering

These are the CLAIMS AND SPECIFICATIONS I-1.
May 27, 2009

Parts Removed to Make: Partial Views

13. FIG. XIII This figure shown here is of the **Partial View**, of the **Front** of Model-X-1 of the fully assembled, Perimeter Saw and its' Components Housing insert Attachment with the center opening for the Component Center Cover. The part of the Blade Housing Insert Attachment that is molded into the Components Housing Center Cover is the Blade Holding Apparatus. This part covers the Blade and it is covered by the Friction Absorbing Pad.

1. Fig. XIII.-a The Fully Assembled Perimeter Saw Reduced in Size. The Perimeter Saw is reduced here in size and the order of assembly here represents the ability to make the product smaller. This process can be done in a larger model unit and the applied to this drawing, Fig. IX-a.

2. Fig. XIII-b The Fully Assembled Perimeter Saw Reduced in Size. The Perimeter Saw is reduced here in size and the order of assembly here represents the ability to make the product smaller. This process can be done in a larger model unit and the applied to this drawing, Fig. IX-a.

Tropical Saws & Export Corporation ™

Perimeter Saw © ®

Application Drawings and Prints
Joe Nathan Brown-Inventor, Designer, Author, Engineering

These are the CLAIMS & SPEICIFICATIONS I-1.

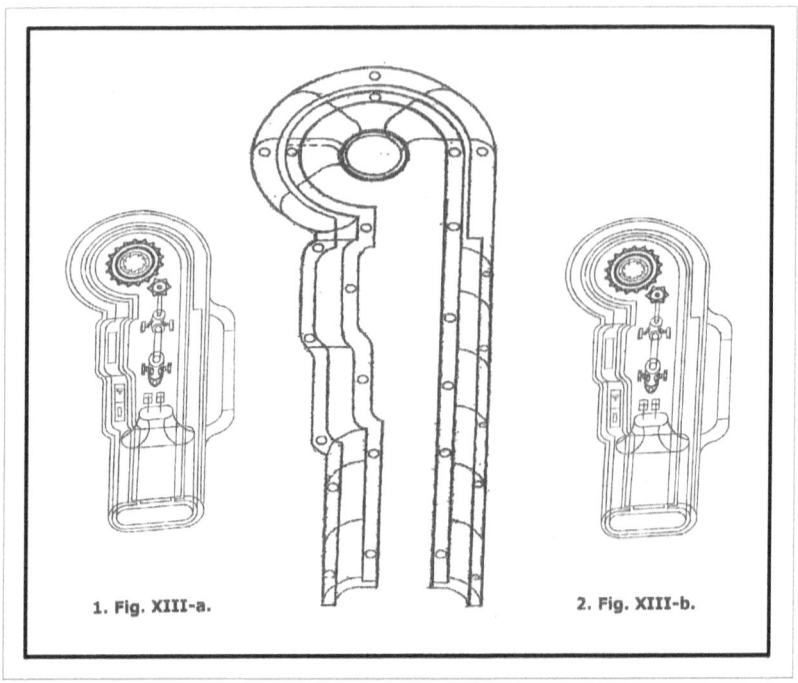

1. Fig. XIII-a.   2. Fig. XIII-b.

13. FIG. XIII.

13. FIG. XIII. This figure shown here is of the, **Partial View** of the **Front of Model X-1** of the fully assembled **Perimeter Saw,** pages **42** to **51**. The parts are described in detail on **p**age **46** of the Claims and Specifications.

Page 47. of 140.

Tropical Saws & Export Corporation ™

Perimeter **S**aw-**D**esign **P**atent **A**pplication
Joe Nathan Brown-Inventor, Designer, Author, Engineering

These are the CLAIMS AND SPECIFICATIONS I-1.
February 23, 2009

Parts Removed to Make: Partial Views

14. FIG. XIV This figure shown here is of the **Partial View**, of the **Front** of Model-X-1 of the fully assembled, Perimeter Saw and its' Frontal Components Center Cover.

1. Fig. XIV.-a The Components Insert Attachment, Center Cover.

2. Fig. XIV-b The Components Insert Attachment, Center Cover.

14.1. The Components Insert Attachment, Center Cover is the part of the saw that is responsible for the Blade staying in perfect rotation, by the application of the Blade Holding Apparatus in its' top upper region and location, Fig. XXIV-a

14.2. The Component Insert Attachment, Center Cover has in its' top region and section the Blade, Friction Absorbing Pad and since this application is on the front of the blade it does not get a Blade Stability Circle. This piece is made to reach down onto the front of the Blade in perfect alignment to cause a strong fit and symmetric hold.

Page 48. of 140.

Tropical Saws & Export Corporation ™

Perimeter Saw © ®

Application Drawings and Prints
Joe Nathan Brown-Inventor, Designer, Author, Engineering

These are the **CLAIMS & SPEICIFICATIONS I-1.**

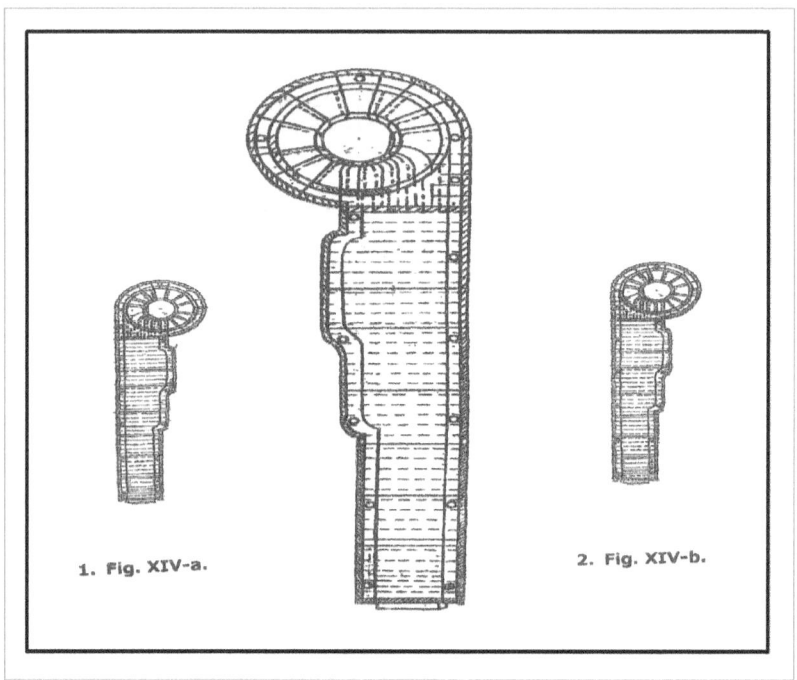

1. Fig. XIV-a.

2. Fig. XIV-b.

14. FIG. XIV.

14. FIG. XIV. This figure shown here is of the**, Partial View** of the **Front of Model X-1** of the fully assembled **Perimeter Saw,** pages **42** to **51**. The parts are described in detail on **p**age **48** of the Claims and Specifications.

Page 49. of 140.

Tropical Saws & Export Corporation ™

Perimeter Saw-Design Patent Application
Joe Nathan Brown-Inventor, Designer, Author, Engineering

These are the CLAIMS AND SPECIFICATIONS I-1.
February 23, 2009

Parts Removed to Make: Partial Views

15. FIG. XV.   This figure shown here is of the **Partial View**, a **Rear.**

This is the one of the Views, showing Model-X-1 of the fully assembled, Perimeter saw as shown here and throughout the book. It shows the Components Housing Insert Attachment, facing downward to show the The figures shown here is of the difference in it and the front Components Housing Insert Attachment.

1. Fig. XV.-a The Fully Assembled, Perimeter Saw Rear View. The Perimeter

Saw is reduced here in size and the order of assembly here represents. The

ability to make the product smaller, adds to more products within this saw's capacity. This process can be done in a larger Model X-1 unit and is applied to this drawing, Fig. IX-a.

2. Fig. XV-b   This is the Fully Assembled, Perimeter Saw Rear View. The

Perimeter Saw is reduced here in size and the order of assembly here.

This illustration represents the ability to make the product smaller, because it can be a customed built one for this purpose. This process can be.

It is done in a larger model unit and the applied to this drawing, Fig. IX-a.

Tropical Saws & Export Corporation ™

Perimeter Saw © ®

Application Drawings and Prints
Joe Nathan Brown-Inventor, Designer, Author, Engineering

These are the CLAIMS & SPEICIFICATIONS I-1.

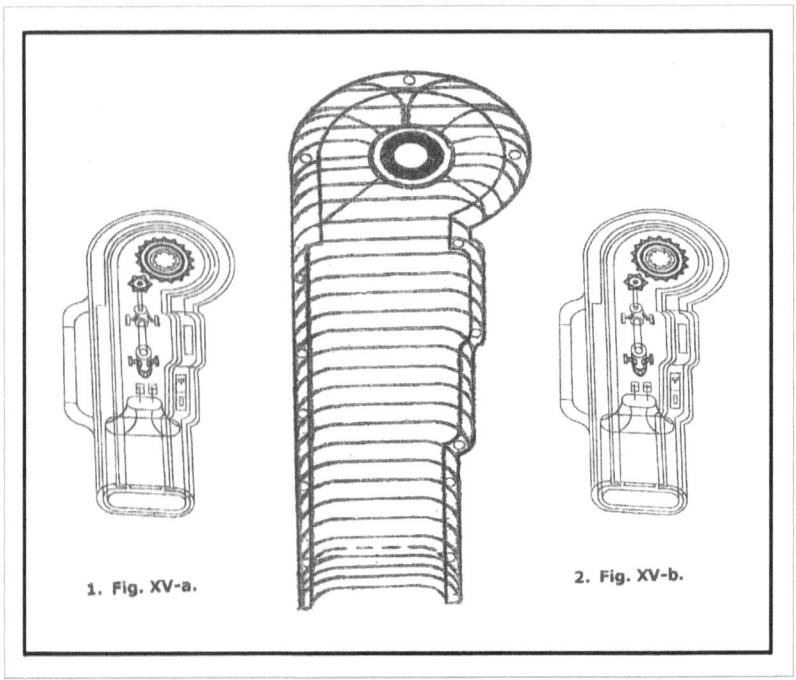

1. Fig. XV-a.

2. Fig. XV-b.

15. FIG. XV.

15. FIG. XV This figure shown here is of the, **Partial View** of the **Front of Model X-1** of the fully assembled **Perimeter Saw,** pages **42** to **51**. The parts are described in detail on **page 50** of the Claims and Specifications.

Page 51. of 140.

Tropical Saws & Export Corporation ™

Perimeter Saw-Design Patent Application
Joe Nathan Brown-Inventor, Designer, Author, Engineering

These are the CLAIMS AND SPECIFICATIONS I-1.
February 23, 2009

Two or More Parts or Segments in an Area: Sectional View

16. FIG. XVI.   This figure shown here is of the **Sectional View**, of the **Top Plan View** of **Model-X-1** of the fully assembled, Perimeter Saw and its' the Blade Holding Region and is a new section.   This is only a piece of the true form that this section This part will have taken its form from the outer edges of the blade.   The picture shows that the top part will be circular in form and consistent with the upper area made.

Having a part engineered to the Blade housing of the blade, for specific cutting reasons is done as an advantage.   The shape that the top has is actually round from the Component Housing Insert Attachment, Front and Rear is the distance being considered in this figure.

Tropical Saws & Export Corporation ™

Perimeter Saw © ®

Application Drawings and Prints
Joe Nathan Brown-Inventor, Designer, Author, Engineering

These are the CLAIMS & SPEICIFICATIONS I-1.

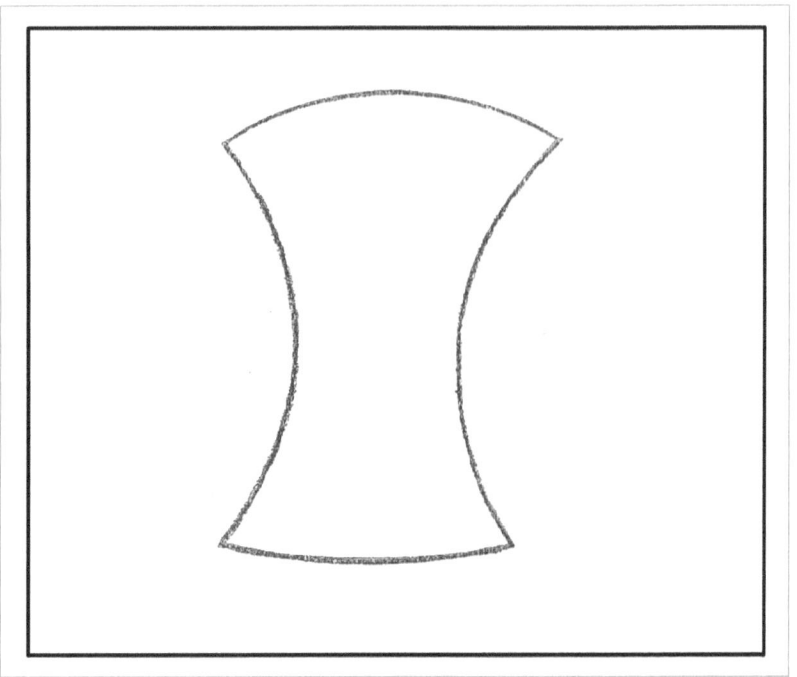

16. FIG. XVI.

16. FIG. XVI This figure shown here is of the **Sectional Views,** of the **Front of Model X-1** of the fully assembled **Perimeter Saw,** pages **52** to **61**. The parts are described in detail on **page 52** of the Claims and Specifications.

Page 53. of 140.

Tropical Saws & Export Corporation ™

Perimeter Saw-Design Patent Application
Joe Nathan Brown-Inventor, Designer, Author, Engineering

These are the CLAIMS AND SPECIFICATIONS I-1.
May 27, 2009

Two or More Parts or Segments in an Area: Sectional View

17. FIG. XVII. This figure shown here is of the **Sectional View**, Model-X of the fully assembled, and Perimeter Saw.

1. Fig. XVII.-a. The piece of the Bottom Components Cover, section on the Frame, will be held together by 4 screws.

2. Fig. XVII-b The Bottom Components Cover, with four screws attachment holes.

17.1. The piece of the Bottom Components Cover, is shown to bring

Attention to the shape it has, where the Component Center Cover fits in the

L shaped space and the 4 holes that are threaded, Fig. XVII-a.

17.2. The Bottom Component Cover shown here, fit on Fig. XVII-a. With the

use of 4 connecting screws, so that it can be easily removed when it is

necessary to change and recharge the Battery.

Tropical Saws & Export Corporation ™

Perimeter Saw © ®

Application Drawings and Prints
Joe Nathan Brown-Inventor, Designer, Author, Engineering
These are the CLAIMS & SPEICIFICATIONS I-1.

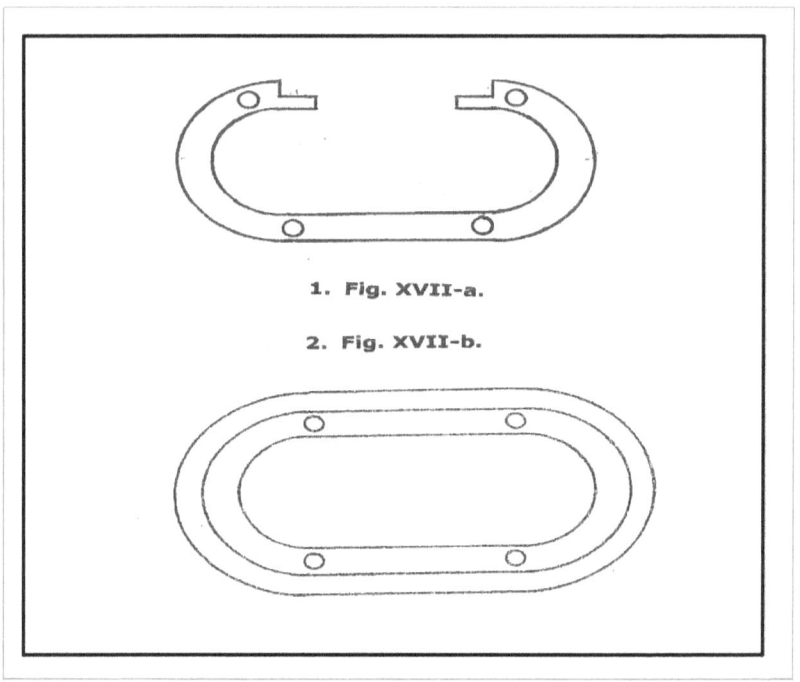

1. **Fig. XVII-a.**

2. **Fig. XVII-b.**

17. FIG. XVII.

17. FIG. XVII. This figure shown here is of the **Sectional Views,** of the **Front of Model X-1** of the fully assembled **Perimeter Saw,** pages **52** to **61.** The parts are described in detail on **p**age **54** of the Claims and Specifications.

Tropical Saws & Export Corporation ™

Perimeter Saw-Design Patent Application
Joe Nathan Brown-Inventor, Designer, Author, Engineering

These are the CLAIMS AND SPECIFICATIONS I-1.
February 23, 2009

Two or More Parts or Segments in an Area: Sectional View

18. FIG. XVIII. This figure shown here is of the **Sectional View**, Model-X

Of the fully assembled, Perimeter Saw, it is describe as follows. The Left side view, having the top, Middle and bottom of it displayed and molded into one piece. In the actual Drawing a line would be drawn in the center of this section from the top to The bottom is the lower section of the handle. This line would show the adjoining part of the Component This is the Blade Housing Insert Attachment, Front and Rear, that is used to hold the blade.

1. Fig. XVIII.-a. The Fully Assembled, Perimeter Saw Rear View. The

Perimeter Saw is reduced here in size and the order of assembly here

Representing the ability to make the product smaller is shown with it in shorter measurements. This process can be.

For example the increase in size is done in a larger model unit and the applied to this drawing, Fig. IX-a.

2. Fig. XVIII-b The Fully Assembled, Perimeter Saw Rear View. The
Perimeter Saw is reduced here in size and the order of assembly here
Represents the ability to make the product smaller, and is for working smaller materials. This process can be.
Done in larger model units the perimeter saw is built and the dimensions are applied to this drawing as is shown in, Fig. IX-a.

Tropical Saws & Export Corporation ™

Perimeter Saw © ®

Application Drawings and Prints
Joe Nathan Brown-Inventor, Designer, Author, Engineering
These are the CLAIMS & SPEICIFICATIONS I-1.

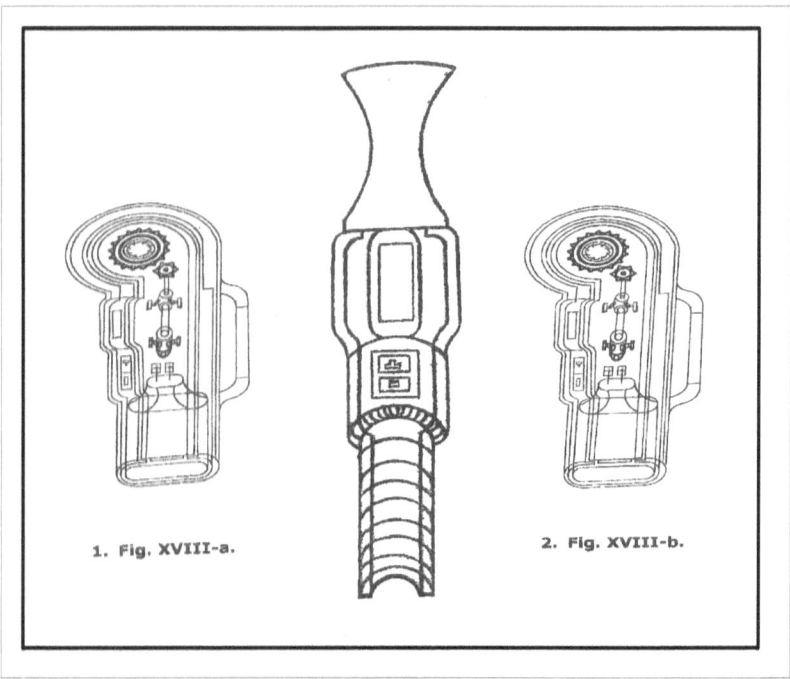

1. Fig. XVIII-a.   2. Fig. XVIII-b.

18. FIG. XVIII.

18. FIG. XVIII. This figure shown here is of the **Sectional Views,** of the **Front of Model X-1** of the fully assembled **Perimeter Saw,** pages **52** to **61**. The parts are described in detail on page **56** of the Claims and Specifications.

Tropical Saws & Export Corporation ™

Perimeter Saw-Design Patent Application
Joe Nathan Brown-Inventor, Designer, Author, Engineering

These are the CLAIMS AND SPECIFICATIONS I-1.
May 27, 2009

Two or More Parts or Segments in an Area: Sectional View

19. FIG. XIX.   This figure shown here is of the **Sectional View**, Model-X
Of the fully assembled, Perimeter Saw.   In this illustration it is prepared for the
preceding illustration. The drawing displays is only a piece. The left side view
giving more perspective to the section it was removed. It is comprised of the
Blade Holding Apparatus, in the upper area. The Push Button Switch space left
open is in clear view.

1. Fig. I.-a. This is the Fully Assembled, Perimeter Saw Rear View. The
Perimeter Saw is reduced here in size and the order of assembly here
represents the Ability to make the product smaller, over again helps presents
the options available. This process can be done in a larger Model unit and then
applied to this drawing. Fig. IX-a.

Copyright registration © ® Joe Nathan Brown 2009

Tropical Saws & Export Corporation ™

Perimeter Saw © ®

Application Drawings and Prints
Joe Nathan Brown-Inventor, Designer, Author, Engineering
These are the CLAIMS & SPEICIFICATIONS I-1.

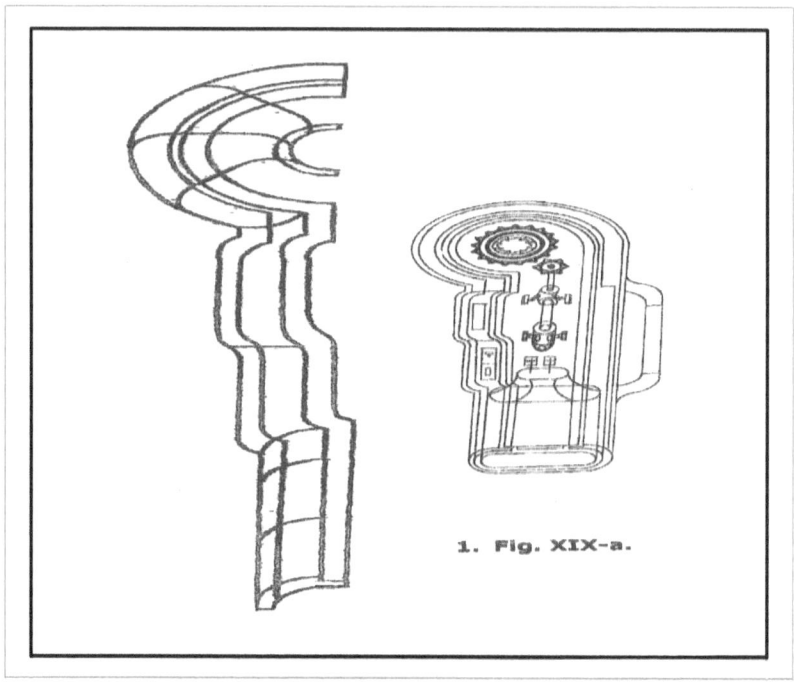

1. Fig. XIX-a.

19. FIG. XIX.

19. FIG. XIX. This figure shown here is of the **Sectional Views,** of the **Front of Model X-1** of the fully assembled **Perimeter Saw,** pages **52** to **61**. The parts are described in detail on **page 58** of the Claims and Specifications.

Page 59. of 140.

Tropical Saws & Export Corporation ™

Perimeter Saw-Design Patent Application
Joe Nathan Brown-Inventor, Designer, Author, Engineering

These are the CLAIMS AND SPECIFICATIONS I-1.
February 23, 2009

Two or More Parts or Segments in an Area: Sectional View

20. FIG. XX.   This figure shown here is of the **Sectional View**, Model-X.

And it is the fully assembled, Perimeter Saw. The picture shows that on the

right. On this side view of the Components Housing Insert Attachment, Front

and Rear there is A small space of separation, where the two parts fit together,

and is the adjoining space. It has on the Bottom region the screw holes in the

location that is best suited for them. And the Bottom Component Cover

will fit over these holes. The handle lower elevation means that the attaching

handle screw sections is not surface mounted.

Page 60. of 140.

Tropical Saws & Export Corporation ™

Perimeter Saw © ®

Application Drawings and Prints
Joe Nathan Brown-Inventor, Designer, Author, Engineering
These are the CLAIMS & SPEICIFICATIONS I-1.

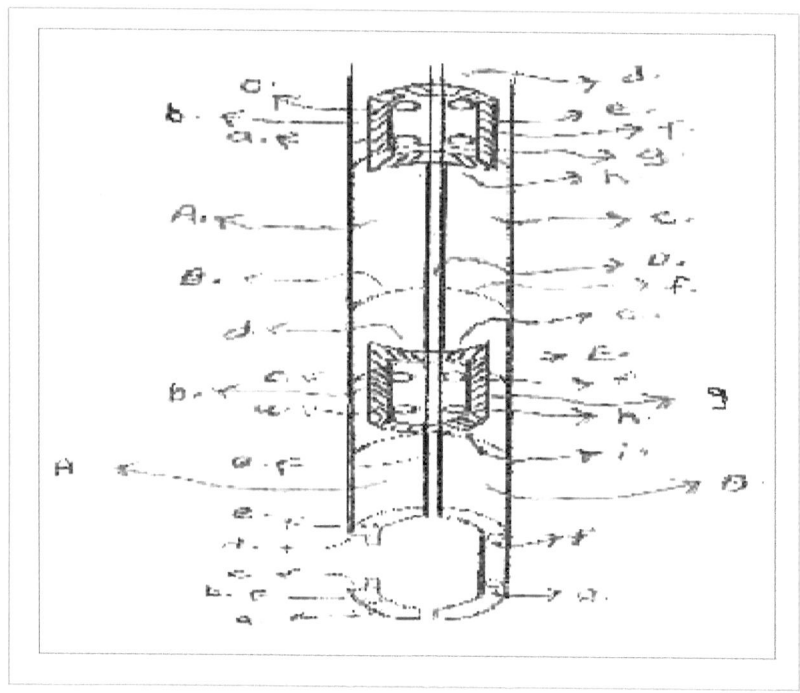

20. FIG. XX.

20. FIG. XX. This figure shown here is of the **Sectional Views,** of the **Front of Model X-1** of the fully assembled **Perimeter Saw,** pages **52** to **61**. The parts are described in detail on **page 60** of the Claims and Specifications.

Page 61. of 140.

Tropical Saws & Export Corporation ™

Perimeter Saw-Design Patent Application
Joe Nathan Brown-Inventor, Designer, Author, Engineering

These are the CLAIMS AND SPECIFICATIONS I-1.
February 23, 2009

The Views rotated and separated: are the Alternate Positional Views.

21. FIG. XXI. This figure shown here is of the **Alternate Positional View,**

Model-X of the fully assembled, Perimeter Saw. This right Alternate

Positional View here, shows the structural look in appearance that the saw

The saw will have. When a piece of the Component Insert Attachment, Front &

Frame and the Components Center Cover are joined together at the top. It

also shows in appearance the shape that the top region will have, as it is two

Circles adjoined by a line, therein making a flat and round forming part.

Tropical Saws & Export Corporation ™

Perimeter Saw © ®

Application Drawings and Prints
Joe Nathan Brown-Inventor, Designer, Author, Engineering
These are the CLAIMS & SPEICIFICATIONS I-1.

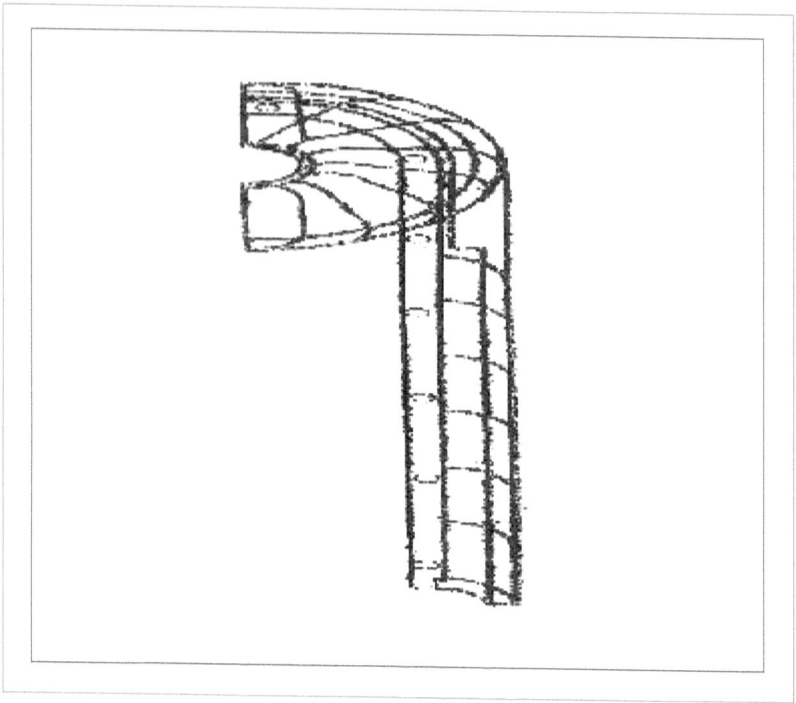

21. FIG. XXI

21. FIG. XXI. This figure shown here is of the **Alternate Positional Views,** of the **Front of Model X-1** of the fully assembled **Perimeter Saw,** pages **62** to **81**. The parts are described in detail on **page 62** of the Claims and Specifications.

Page 63. of 140.

Tropical Saws & Export Corporation ™

Perimeter **S**aw-**D**esign **P**atent **A**pplication
Joe Nathan Brown-Inventor, Designer, Author, Engineering

These are the CLAIMS AND SPECIFICATIONS I-1.
May 27, 2009

Views Rotated and Separated: Alternate Positional Views

22. FIG. XXII This figure shown here is of the **Alternate Positional Forms View**, Model-X of the fully assembled, perimeter Saw. This drawing shows the inner space of the Components Housing Insert Attachment, where the Parts and Piece of the saw will be attached. It shows the depth indication and the raised area made possible by the Circular Form of the Blade Holding Apparatus molded into the Frame. There are a total of 12 screw holes possibilities for mounting the Component Housing Insert Attachment, Front Cover. And 2 more holes to install the Component Bottom Cover.

1. Fig. XXII.-a The Components Housing Insert Attachment in Reverse Angled View.

2. Fig. XXII-b The Components Housing Insert Attachment in Reverse Angled View.

22.1 The Component Housing Insert Attachment in Reverse Angled View, add to the possibilities of allowing the Utility Handle to be placed on the left side of the saw and the switches on the right.

22.2 The Component Housing Insert Attachment in Reverse Angled View, add to the possibilities of allowing the Utility Handle to be placed on the left side of the saw and the switches on the right.

22.3 The Components Mounting Screws-Phillips Head.

22.4 The Components Mounting Screws-Align Wrench Head.

Page 64. of 140.

Tropical Saws & Export Corporation ™

Perimeter Saw © ®

Application Drawings and Prints
Joe Nathan Brown-Inventor, Designer, Author, Engineering
These are the CLAIMS & SPEICIFICATIONS I-1.

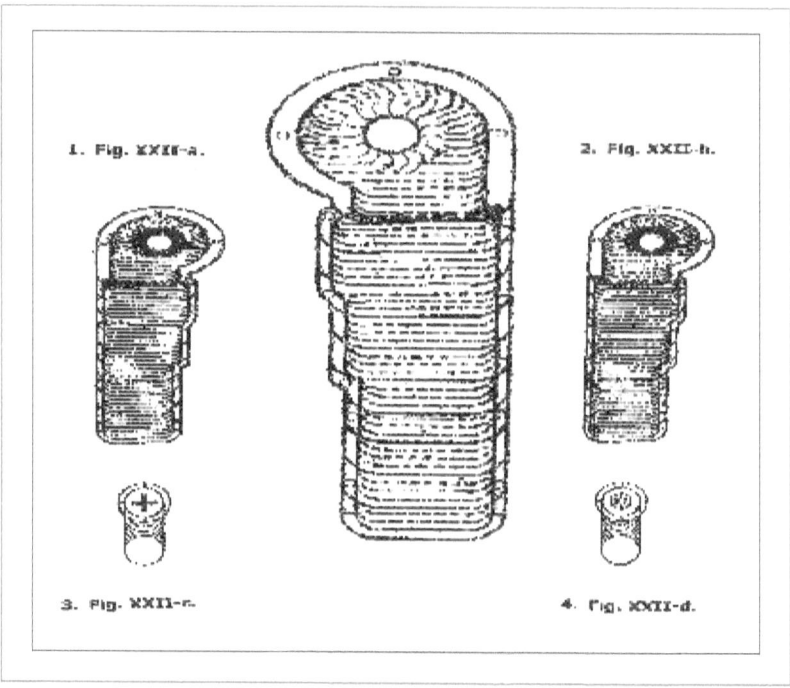

1. Fig. XXII-a.

2. Fig. XXII-h.

3. Fig. XXII-c.

4. Fig. XXII-d.

22. FIG. XXII.

22. FIG. XXII. This figure shown here is of the **Alternate Positional Views,** of the **Front of Model X-1** of the fully assembled **Perimeter Saw,** pages **62 to 81.** The parts are described in detail on page **64** of the Claims and Specifications.

Tropical Saws & Export Corporation ™

Perimeter Saw-Design Patent Application
Joe Nathan Brown-Inventor, Designer, Author, Engineering

These are the CLAIMS AND SPECIFICATIONS I-1.
February 23, 2009

These Views rotated and separated: are the Alternate Positional Views.

23. FIG. XXIII. This figure shown here is of the **Alternate Positional View**, Model-X of the fully assembled, perimeter Saw. This figure is of the Blade with its Cutting Edges in the Center by design and engineering, Stability Circle on the Rear and the Square shaped gear section. The Square shaped Gear section will have a depth of 1/16″, 1/32″ + 1/16″, 1/8″, 1/8″ + 1/32″, 3/16″, 3/16″ + 1/32, and 1/4″ depending on the depth of the blade they are machined in.

1. Fig. XXIII.-a The Circular Blade's Cutting Edges diagram.

2. Fig. XXIII-b The Circular Blade's Cutting Edges, diagram.

23.1. The Blade's Cutting Edges, diagram shown here is to display the way the blade cutting edges are made and in what direction they are place to cut the materials. This type cutting edge will cut the materials in a counter clockwise rotation of the blade. The circular raised area on the teeth is made and designed to direct the discarded materials off and out of the blade space.

23.2. The Blade's Cutting Edges, diagram shown here is to show that the cutting edges are made sharpened to push the materials out and off the blade circular edge.

Page 66. of 140.

Tropical Saws & Export Corporation ™

Perimeter Saw © ®

Application Drawings and Prints
**Joe Nathan Brown-Inventor, Designer, Author, Engineering
These are the CLAIMS & SPEICIFICATIONS I-1.**

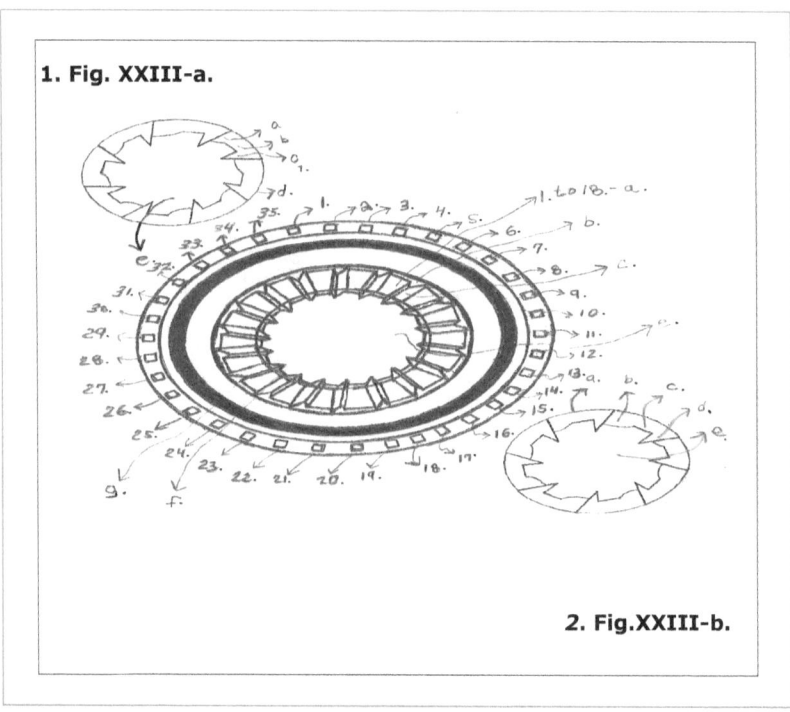

1. Fig. XXIII-a.

2. Fig.XXIII-b.

23. FIG. XXIII.

23. FIG. XXIII This figure shown here is of the **Alternate Positional Views,** of the **Front of Model X-1** of the fully assembled **Perimeter Saw,** pages **62 to 81.** The parts are described in detail on page **66** of the Claims and Specifications.

Page 67. of 140.

Tropical Saws & Export Corporation ™

Perimeter **S**aw-**D**esign **P**atent **A**pplication
Joe Nathan Brown-Inventor, Designer, Author, Engineering

These are the CLAIMS AND SPECIFICATIONS I-1.

**These Views rotated and separated: are the Alternate Positional
Views.**

24. FIG. XXIV This figure shown here is of the **Alternate Positional View**,
and they are Model-X of the fully assembled, Perimeter Saw.

1. Fig. XXIV.-a is of the Circular Blade Counter Clockwise Cutting Edges, as
this diagram shows.

2. Fig. XXIV-b The Circular Blade Counter Clockwise Cutting Edges, Diagram.

24.1. The Circular Blade's Cutting Edges, diagram is presented here for the
observation of the ability to put teeth in the center of a circle and have them
engineered in perfect alignment for circular rotation of the blade.   So as to
cause an equal amount of pressure to be applied to the materials with the
movement of the hand to determined the material separation, while making a
cut.   These diagram shows a counter clockwise rotation of the blade and is Cut
1-A.   And the reason for this is that the motor would have to turn its shaft in
this manor and be label as the forward movement, applied to the
Forward/Reverse switch capabilities.   The purpose of the turning action is that
the holding of the handle be griped, because the saw would turn in the same
direction of the cutting edges if held loosely.

24.2.   The Circular Blade's Cutting Edges, diagram is presented here for the
observation of the ability to put teeth in the center of a circle an have
them engineered in perfect alignment for circular rotation of the blade.

Tropical Saws & Export Corporation ™

Perimeter Saw © ®

Application Drawings and Prints
Joe Nathan Brown-Inventor, Designer, Author, Engineering
These are the CLAIMS & SPEICIFICATIONS I-1.

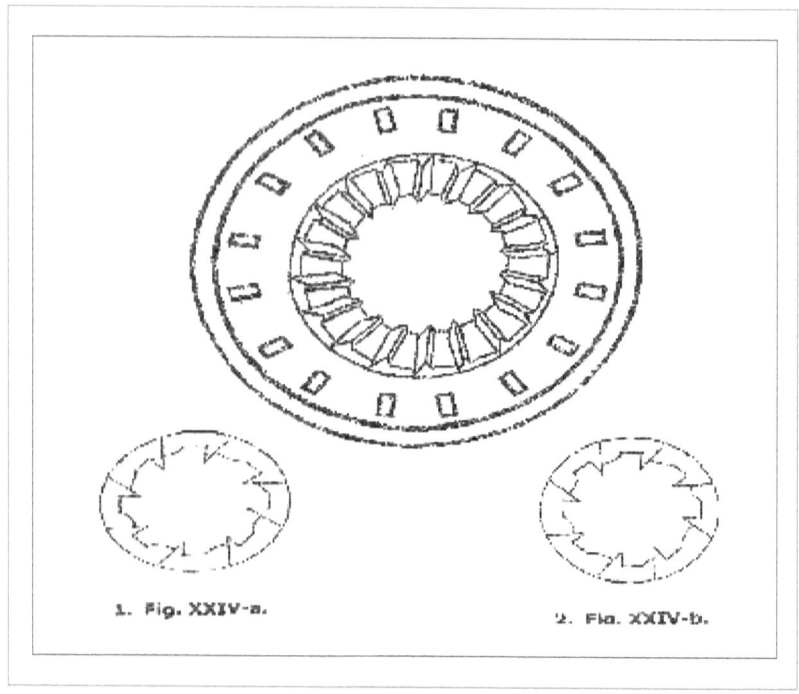

1. Fig. XXIV-a.

2. Fig. XXIV-b.

24. FIG. XXIV.

24. FIG. XXIV This figure shown here is of the **Alternate Positional Views**, of the Front of Model X-1 of the fully assembled **Perimeter Saw**, pages **62 to 81**. The parts are described in detail on page **68** of the Claims and Specifications.

Page 69. of 140.

Tropical Saws & Export Corporation ™

Perimeter Saw-Design Patent Application
Joe Nathan Brown-Inventor, Designer, Author, Engineering

These are the CLAIMS AND SPECIFICATIONS I-1.
February 23, 2009

These Views Rotated and Separated: are the Alternate Positional Views

25. FIG. XXV. . This figure shown here is of the **Alternate Positional View**, Model-X of the fully assembled, and Perimeter Saw.

1. Fig. XXV.-a is shown to show the difference in the Circular Blade Counter Clockwise Cutting Edges, Diagram.

2. Fig. XXV-b The Circular Blade Counter Clockwise Cutting Edges, Diagram.

25.1. The Circular Blade's Cutting Edges, diagram is presented here for the observation of the ability to put teeth in the center of a circle a have them engineered in perfect alignment for circular rotation of the blade. So as to cause that an equal amount of pressure to be applied to the materials with the movement of the hand to determined the material separation, while making a cut. These diagram shows a counter clockwise rotation of the blade and is Cut # 1-A. And the reason for this is that the motor would have to turn its shaft in that manor and be label as the forward movement, applied to the Forward/Reverse switch capabilities. The purpose of the turning action is that the holding of the handle be griped, because the saw would turn in the same direction of the cutting edges if held loosely.

25.2. The Blade's Cutting Edges, diagram shown here is to show that the cutting edges are made sharpened to push the materials out and off the blade circular edge.

Tropical Saws & Export Corporation ™

Perimeter Saw © ®
Application Drawings and Prints
Joe Nathan Brown-Inventor, Designer, Author, Engineering

These are the CLAIMS AND SPECIFICATIONS I-1.

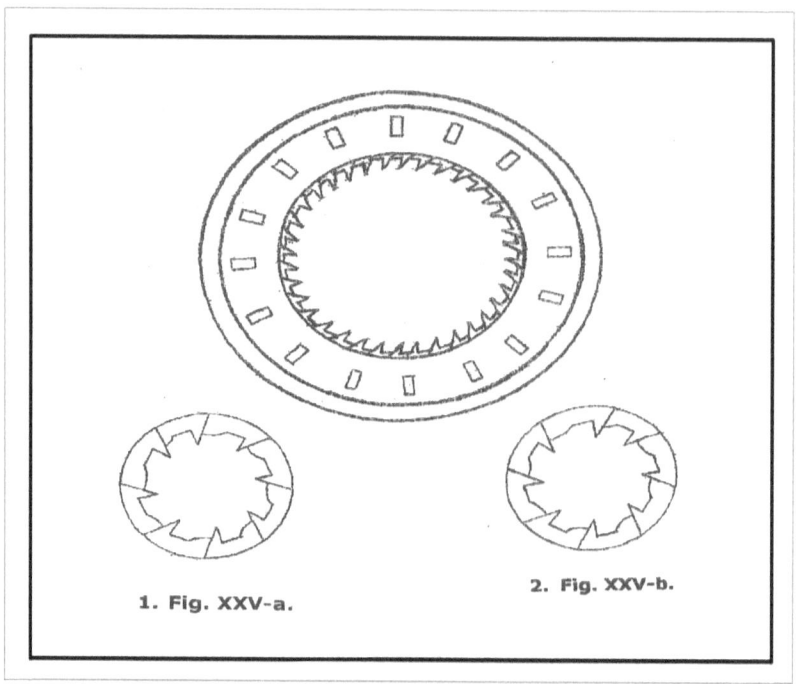

1. Fig. XXV-a.

2. Fig. XXV-b.

25. FIG. XXV.

25. FIG. XXV. This figure shown here is of the **Alternate Positional Views**, of the Front of Model X-1 of the fully assembled Perimeter Saw, pages 62 to 81. The parts are described in detail on page **70** of the Claims and Specifications.

Page 71. of 140.

Tropical Saws & Export Corporation ™

Perimeter Saw-Design Patent Application
Joe Nathan Brown-Inventor, Designer, Author, Engineering

These are the CLAIMS AND SPECIFICATIONS I-1.

Views Rotated and Separated: Alternate Positional Views

26. FIG. XXVI This figure shown here is of the **Alternate Positional View**, Model-X of the fully assembled, and Perimeter Saw.

1. Fig. XXVI.-a is the Circular Blade with the square gear holes;
2. Fig. XXVI-b Is the Left Side View of the Circular Blade;
3. Fig. XXVI-c Is the Top Plan View of the Circular Blade;
4. Fig. XXVI-d Is the Right Side View of the Circular Blade;
5. Fig. XXVI-e Is the Gear space of the square gear holes;
 6. Fig. XXVI-f Is the Bottom Plan View of the Circular Blade;

7. Fig. XXVI-g Is the Blade Stability Circle & Cutting Edges;

26.1. The Circular Blade with the square gear holes, are in the center of the page. It is Blade type 2 second to the external Gear Activator levers.

26.2. The Left Side View of the Circular Blade, shown here to indicate the circular form of the blade.

26.3. The Top Plan View of the Circular Blade, shown here to indicate the circular form of the blade.

26.4. The Right Side View of the Circular Blade, shown here to indicate the circular form of the blade.

26.5. The Gear space of the square gear holes, shown here as a separate piece of the blade.

26.6. The Bottom Plan View of the Circular Blade, shown here to indicate the circular form of the blade.

26.7. The Blade Stability Circle & Cutting Edges, as two separate pieces that are united in the assembling of the blade.

Tropical Saws & Export Corporation ™

Perimeter Saw © ®
Applications Drawings and Blue Prints
Joe Nathan Brown-Inventor, Designer, Arthur, Engineering

These are the CLAIMS AND SPECIFICATIONS I-1.

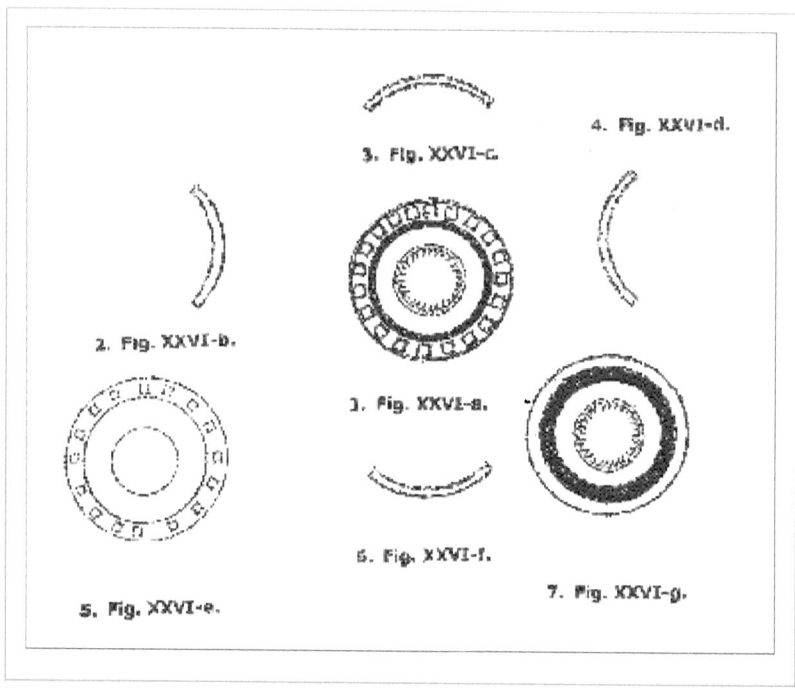

4. Fig. XXVI-d.

3. Fig. XXVI-c.

2. Fig. XXVI-b.

1. Fig. XXVI-a.

5. Fig. XXVI-e.

6. Fig. XXVI-f.

7. Fig. XXVI-g.

26. FIG. XXVI.

26. FIG. XXVI This figure shown here is of the **Alternate Positional Views,** of the **front on Model X-1,** of the fully assembled **Perimeter Saw,** pages **62** to **81.** The parts are described in details on page **72** of the Claims and Specifications.

Page 73. of 140.

Tropical Saws & Export Corporation ™

Perimeter Saw-Design Patent Application
Joe Nathan Brown-Inventor, Designer, Author, Engineering

These are the CLAIMS AND SPECIFICATIONS I-1.
May 27, 2009

Views Rotated and Separated: Alternate Positional Views

27. FIG. XXVII This figure shown here is of the **Alternate Positional View**, Model-X of the fully assembled, and Perimeter Saw.

1. **Fig. XXVII-a**   The Blades' full circular View of the outer diameter;
2. **Fig. XXVII-b**   The Blades' Left Side View of the outer diameter;
3. **Fig. XXVII-c**   The Blades' Top Plan View of the outer diameter;
4. **Fig. XXVII-d**   The Blades' Right Side View of the outer diameter;
5. **Fig. XXVII-e**   The Blades' Stability Circle on the rear View;
6. **Fig. XXVII-f**   The Blades' Bottom Plan View of the outer diameter;

27.1. The Blades' full circular View of the outer diameter, shown here details the form of the blade being a circle with a 1/16" to 1/2" inch depth.

27.2. The Blades' Left Side View of the outer diameter, shown here details the form of the blade with a look at a piece with a length of 90 degrees.

27.3. The Blades' Top Plan View of the outer diameter, shown here details the form of the blade with a look at a piece with a length of 90 degrees.

27.4. The Blades' Right Side View of the outer diameter, shown here details the form of the blade with a look at a piece with a length of 90 degrees.

27.5. The Blades' Stability Circle on the rear View, shown here tell the actual location of the Circle on the blade, leaving space in the radius.

27.6. The Blades' Bottom Plan View of the outer diameter, shown here details the form of the blade with a look at a piece with a length of 90 degrees.

Tropical Saws & Export Corporation ™

Perimeter Saw © ®

Application Drawings and Prints
Joe Nathan Brown-Inventor, Designer, Author, Engineering
These are the CLAIMS AND SPECIFICATIONS I-1.

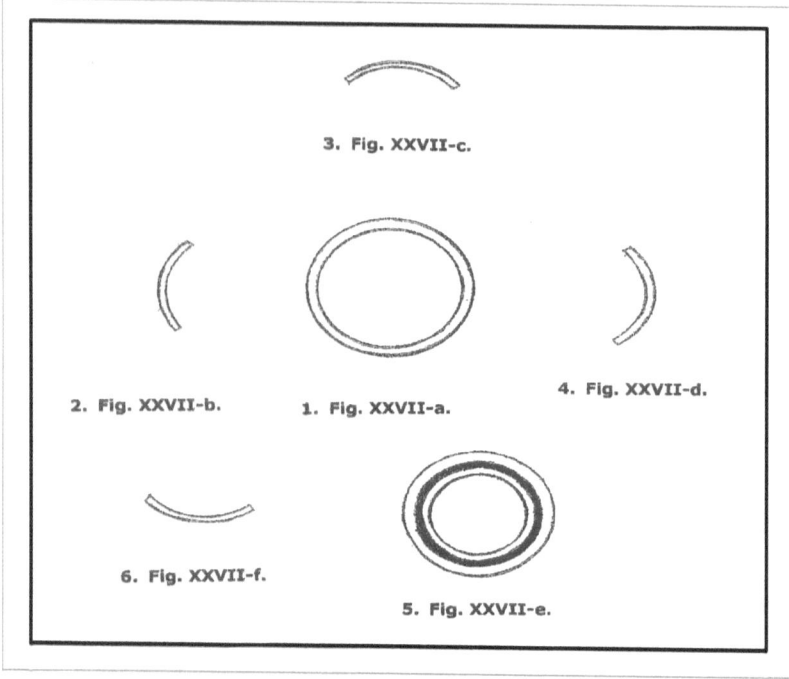

3. Fig. XXVII-c.

2. Fig. XXVII-b.   1. Fig. XXVII-a.   4. Fig. XXVII-d.

6. Fig. XXVII-f.

5. Fig. XXVII-e.

27. FIG. XXVII.

27. FIG. XXVII. This figure shown here is of the **Alternate Positional View,** of the **Front of Model X-1** of the fully assembled **Perimeter Saw,** pages **62** to **81**. The parts are described in details on page **74** of the Claims and Specifications.

Page 75. of 140.

Tropical Saws & Export Corporation ™

Perimeter Saw-Design Patent Application
Joe Nathan Brown-Inventor, Designer, Author, Engineering

These are the CLAIMS AND SPECIFICATIONS I-1.
February 23, 2009

These Views rotated and separated: are the Alternate Positional Views.

28. FIG. XXVIII. . This figure shown here, is of the **Alternate Positional View**, Model-X, of the fully assembled, Perimeter Saw. It is a drawing of the actual Blade type, which will be used in this Model-x-1 saw, are in this figure. This is a rear view. Views of the blade and on the rear view is the Blade Stability Circle. The Location of the Stability Circle in this picture is an alternate positional view, in the location as it will be, where it is best suited. The Stability circle will Be place where the maximum percentage of radius is left on the X, Y, and Z Axis, that is beyond the Circle. The Stability Circle is on the Rear of the Blade, and it is directly under the Blade Holding Apparatus; which is placed on the front. It is designed to keep the blade in the frame and hold it in place with soundness.

Tropical Saws & Export Corporation ™

Perimeter Saw © ®

Application Drawings and Prints
Joe Nathan Brown-Inventor, Designer, Author, Engineering
These are the CLAIMS AND SPECIFICATIONS I-1.

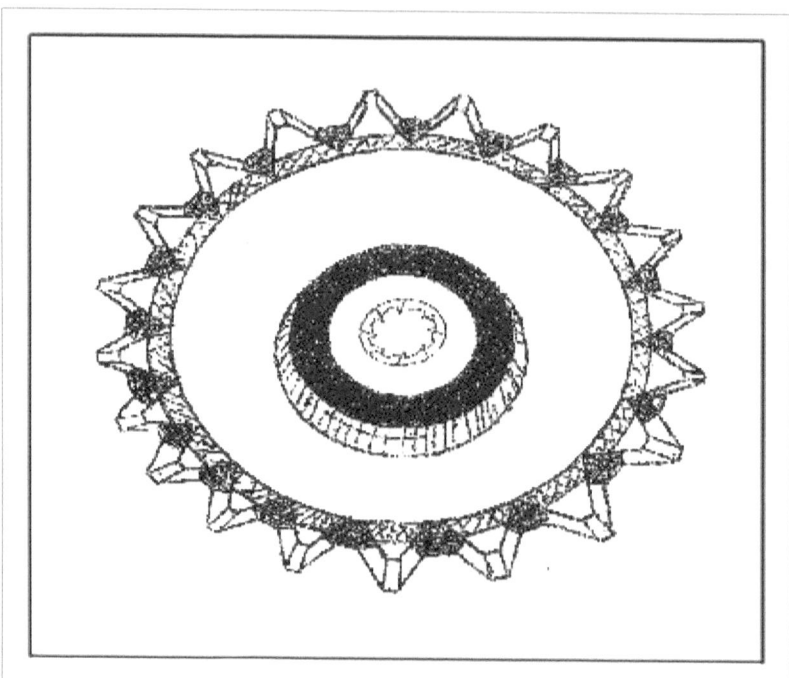

28. FIG. XXVIII.

28. FIG. XXVIII. This figure shown here is of the **Alternate Positional View,** of the **Front of Model X-1** of the fully assembled **Perimeter Saw,** pages **62** to **81**. The parts are described in details on page **76** of the Claims and Specifications.

Page 77. of 140.

Tropical Saws & Export Corporation ™

Perimeter Saw-Design Patent Application
Joe Nathan Brown-Inventor, Designer, Author, Engineering

These are the CLAIMS AND SPECIFICATIONS I-1.
February 23, 2009

Views Rotated and Separated: Alternate Positional Views

29. FIG. XXIX This figure shown here is of the **Alternate Positional View**, Model-X of the fully assembled, and Perimeter Saw.

1. Fig. XXIX.-a is of the Front View of the Blade, its' Gear, Activators, Levers, and Cutting Edges.

2. Fig. XXIX-b is of the Rear View of the Blade, its' Stability Circle and Cutting Edges.

29.1. The Front View of the Blade, its' Gear, Activators, Levers, and Cutting Edges, shown here has the gears made to cause the blade to rotate forward and Reverse; which is counter clockwise and clockwise respectively.

29.2. The Rear View of the Blade, its' Stability Circle and Cutting Edges, shown here will be the way the part of the blade rear side looks in comparison to the front side.

Page 78. of 140.

Copyright registration © ® Joe Nathan Brown 2009

Tropical Saws & Export Corporation ™

Perimeter Saw © ®

Application Drawings and Prints
Joe Nathan Brown-Inventor, Designer, Author, Engineering
These are the CLAIMS AND SPECIFICATIONS I-1.

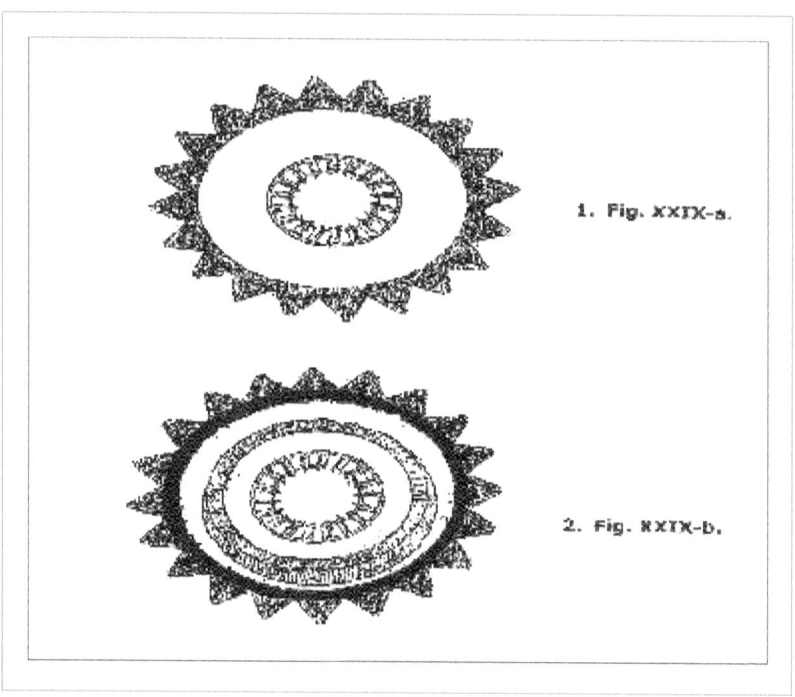

1. Fig. XXIX-a.

2. Fig. XXIX-b.

29. FIG. XXIX.

29. FIG. XXIX This figure shown here is of the **Alternate Positional View,** of the **Front of Model X-1** of the fully assembled **Perimeter Saw,** pages **62** to **81**. The parts are described in details on page **78** of the Claims and Specifications.

Page 79. of 140.

Tropical Saws & Export Corporation ™

Perimeter Saw-Design Patent Application
Joe Nathan Brown-Inventor, Designer, Author, Engineering

These are the CLAIMS AND SPECIFICATIONS I-1.
May 27, 2009

Views Rotated and Separated: Alternate Positional Views

30. FIG. XXX. This figure shown here is of the **Alternate Positional View**, Model-X of the fully assembled, and Perimeter Saw.

1. Fig. XXX.-a. is of the Rear View of the Blade, its' Stability Circle and Cutting Edges.

2. Fig. XXX-b is of the Front View of the Blade, its' Gear, Activators, Levers, and Cutting Edges.

30.1. The Rear View of the Blade, its' Stability Circle and Cutting Edges, shown here has the gears made to cause the blade to rotate forward and Reverse; which is counter clockwise and clockwise respectively.

30.2. The Front View of the Blade, its' Gear, Activators, Levers, and Cutting Edges, shown here will be the way the part of the blade rear side looks in comparison to the front side.

Tropical Saws & Export Corporation ™

Perimeter Saw © ®

Application Drawings and Prints
Joe Nathan Brown-Inventor, Designer, Author, Engineering
These are the CLAIMS AND SPECIFICATIONS I-1.

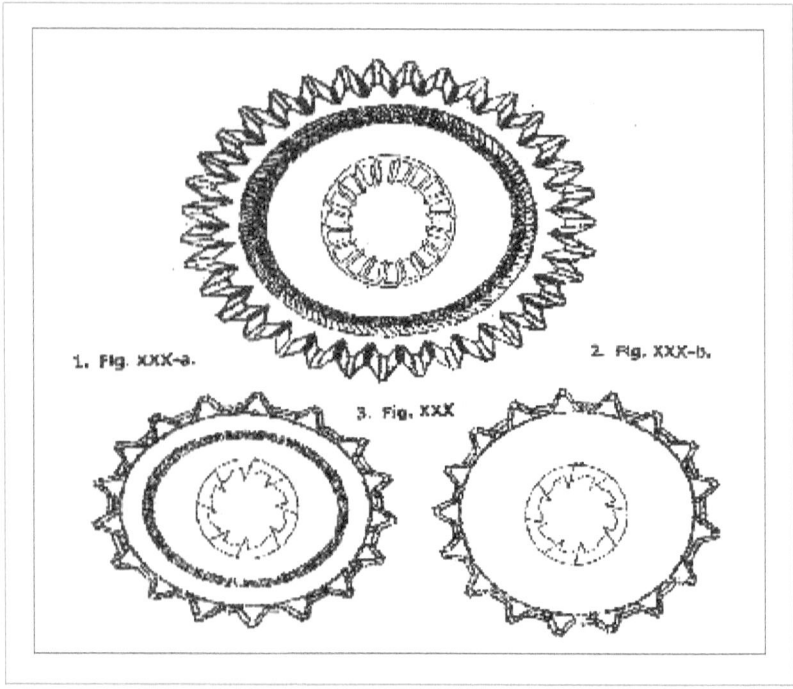

1. Fig. XXX-a.   2. Fig. XXX-b.

3. Fig. XXX

30. FIG. XXX.

30. FIG. XXX This figure shown here is of the **Alternate Positional View,** of the **Front of Model X-1** of the fully assembled **Perimeter Saw,** pages **62** to **81**. The parts are described in details on page **80** of the Claims and Specifications.

Page 81. of 140.

Tropical Saws & Export Corporation ™

Perimeter Saw-Design Patent Application
Joe Nathan Brown-Inventor, Designer, Author, Engineering

These are the CLAIMS AND SPECIFICATIONS I-1.
February 23, 2009

Parts Capabilities Identified and Named: Modified Views

31. FIG. XXXI This figure shown here is of the **Modified Forms View**, Model-X of the fully assembled, and Perimeter Saw.

1. Fig. XXXI.-a is of the Blade with its' Gear Activators & Levers, and Cutting Edges, 31.1. The full circular forms of these gears are here on this front view.

2. Fig. XXXI-b is of the Blades' Left Side of the Blade Gear Activators & Levers, 31.2. This is the piece of the blade covering a 90 degree percentage of the gears.

3. Fig. XXXI-c are of the Blades' Top Plan View of its' Gear Activators & Levers, 31.3. This is the piece of the blade covering a 90 degree percentage of the gears.

4. Fig. XXXI-d is of the Blades' Right Side View of its' Gear Activators & Levers, 31.4 the piece of the blade covering a 90 degree percentage of the gears.

5. Fig. XXXI.-e The Blades' Bottom View of its' Gear Activators & Levers, 31.5. This is the piece of the blade covering a 90 degree percentage of the gears.

6. Fig. XXXI.-f The Blades with a Stability Circle, on the rear, cutting edges and Gear Activators, 31.6. This is the entire look of the blade in a rear view, showing how close to the gear that the blade stability circle is.

Tropical Saws & Export Corporation ™

Perimeter Saw © ®

Application Drawings and Prints
Joe Nathan Brown-Inventor, Designer, Author, Engineering
These are the CLAIMS AND SPECIFICATIONS I-1.

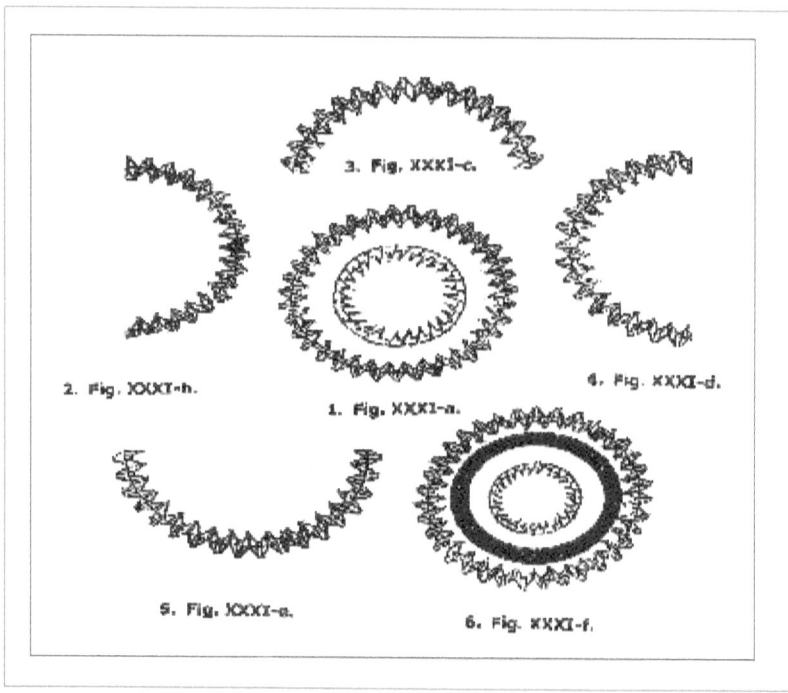

3. Fig. XXXI-c.

2. Fig. XXXI-h.

4. Fig. XXXI-d.

1. Fig. XXXI-a.

5. Fig. XXXI-e.

6. Fig. XXXI-f.

31. FIG. XXXI.

31. FIG. XXXI. This figure shown here is of the **Modified Forms Views,** of the **Front of Model X-1** of the fully assembled **Perimeter Saw,** pages **82** to **101**. The parts are described in details on page **82** of the Claims and Specifications.

Page 83. of 140.

Tropical Saws & Export Corporation ™

Perimeter Saw-Design Patent Application
Joe Nathan Brown-Inventor, Designer, Author, Engineering

These are the CLAIMS AND SPECIFICATIONS I-1.
February 23, 2009

32. FIG. XXXII. This figure shown here is of the **Modified Forms View**, Model-X of the fully assembled, and Perimeter Saw.

1. Fig. XXXII.-a are of the circular blade counter clockwise Cutting Edges, and its' diagram,

2. Fig. XXXII-b The circular blade counter clockwise Cutting Edges, Diagram.

32.1. The Circular Blade's Cutting Edges, diagram is presented here for the observation of the ability to put teeth in the center of a circle a have them engineered in perfect alignment for circular rotation of the blade. So as to cause that an equal amount of pressure to be applied to the materials with the movement of the hand to determined the material separation, while making a cut. These diagram shows a counter clockwise rotation of the blade and is Cut # 1-A. And the reason for this is that the motor would have to turn its shaft in that manor and be label as the forward movement, applied to the Forward/Reverse switch capabilities. The purpose of the turning action is that the holding of the handle be griped, because the saw would turn in the same direction of the cutting edges if held loosely.

32.2. The Circular Blade's Cutting Edges, diagram is presented here for the observation of the ability to put teeth in the center of a circle an have them engineered in perfect alignment for circular rotation of the blade.

Tropical Saws & Export Corporation ™

Perimeter Saw © ®

Application Drawings and Prints
Joe Nathan Brown-Inventor, Designer, Author, Engineering
These are the CLAIMS AND SPECIFICATIONS I-1.

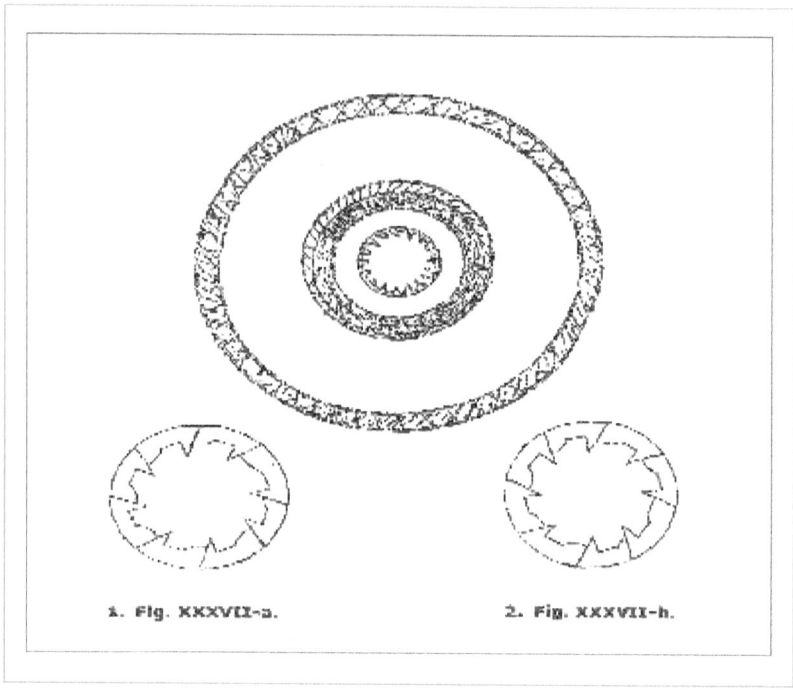

1. Fig. XXXVII-a. 2. Fig. XXXVII-h.

32. FIG. XXXII.

32. FIG. XXXII This figure shown here is of the **Modified Forms Views,** of the **Front of Model X-1** of the fully assembled **Perimeter Saw,** pages **82** to **101.** The parts are described in details on page **84** of the claims and specifications.

Perimeter Saw-Design Patent Application
Joe Nathan Brown-Inventor, Designer, Author, Engineering

These are the CLAIMS AND SPECIFICATIONS I-1.
May 27, 2009

33. FIG. XXXIII This figure shown here is of the **Modified Forms Views,**

of the fully assembled, Perimeter Saw, Model X-1. This figure has the

Blade the Cutting part of the Saw modified to be used on materials that are at,

reach able distances from the outside edge of the saw. And in a position that

is, work able. The new and different blade has two cutting edges. They are

placed. The cutting edges being one above the other for, a double cutting ability

engineer to specification. And this is on this bottom view. The Blade Stability

Circle is still located on the rear section. The Gear Activator, Are Levers that

touches the gear on the Elbow Gear, they have been given space. Between

them; they are designed to facilitate the ease in the rotation of them. The

Blades' gears, when making contact with metal, of the other part, will be. The

parts are precise fitting leaving oil space for lubrication as a heat cooling

system.

Tropical Saws & Export Corporation ™

Perimeter Saw © ®

Application Drawings and Prints
Joe Nathan Brown-Inventor, Designer, Author, Engineering
These are the CLAIMS AND SPECIFICATIONS I-1.

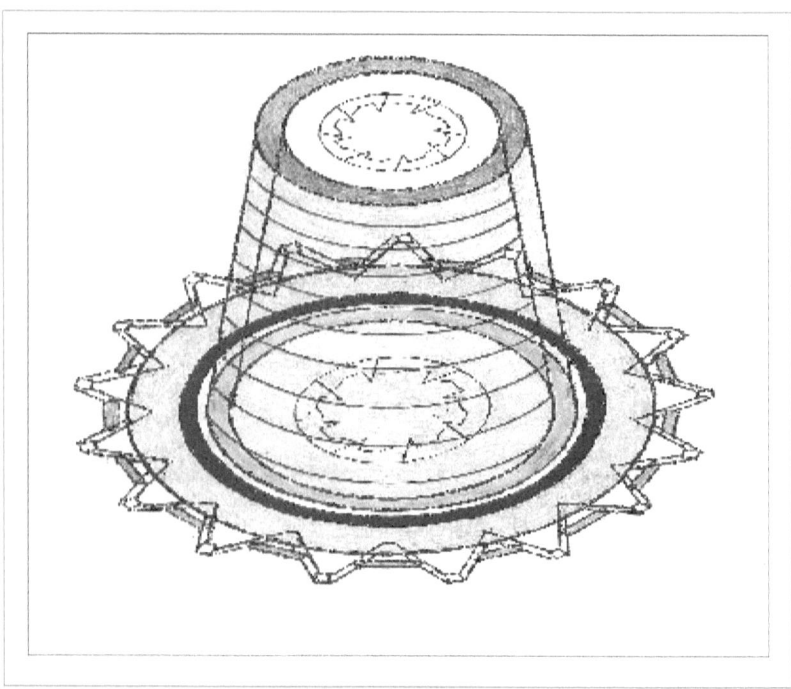

33. FIG. XXXIII.

33. FIG. XXXIII This figure shown here is of the **Modified Forms Views**, of the **Front of Model X-1** of the fully assembled **Perimeter Saw**, pages **82** to **101**. The parts are described in details on page **86** of the claims and specifications.

Tropical Saws & Export Corporation ™

Perimeter Saw-Design Patent Application
Joe Nathan Brown-Inventor, Designer, Author, Engineering

These are the CLAIMS AND SPECIFICATIONS I-1.

34. FIG. XXXIV. . This figure shown here is of the **Modified Forms View**, Model-X of the fully assembled, and Perimeter Saw.

1. Fig. XXXIV.-a is of the Blades' with its' Cutting Edges;

2. Fig. XXXIV-b is of the Blades' Left Side View of the outer diameter;

3. Fig. XXXIV-c is of the Blades' Top Plan View of the outer diameter;

4. Fig. XXXIV-d is of the Right Side View of the Circular Blade;

5. Fig. XXXIV-e is of the Blades' Stability Circle on the rear View;

6. Fig. XXXIV-f is of the Blades' Bottom Plan View of the outer diameter;

34.1. The Blades' with its' Cutting Edges, the Blades' full circular View of the outer diameter, shown here details the form of the blade being a circle with a 1/16" to 1/2" inch depth.

34.2. The Blades' Left Side View of the outer diameter, the Blades' Left Side View of the outer diameter, shown here details the form of the blade with a look at a piece with a length of 90 degrees.

34.3. The Blades' Top Plan View of the outer diameter, the Top Plan View of the Circular Blade, shown here to indicate the circular form of the blade.

34.4. The Right Side View of the Circular Blade, the Blades' Right Side View of the outer diameter, shown here details the form of the blade with a look at a piece with a length of 90 degrees.

34.5. The Blades' Bottom Plan View of the outer diameter, the Blades' Bottom Plan View of the outer diameter, shown here details the form of the blade with a look at a piece with a length of 90 degrees.

Tropical Saws & Export Corporation ™

Perimeter Saw © ®

Application Drawings and Prints
Joe Nathan Brown-Inventor, Designer, Author, Engineering
These are the CLAIMS AND SPECIFICATIONS I-1.

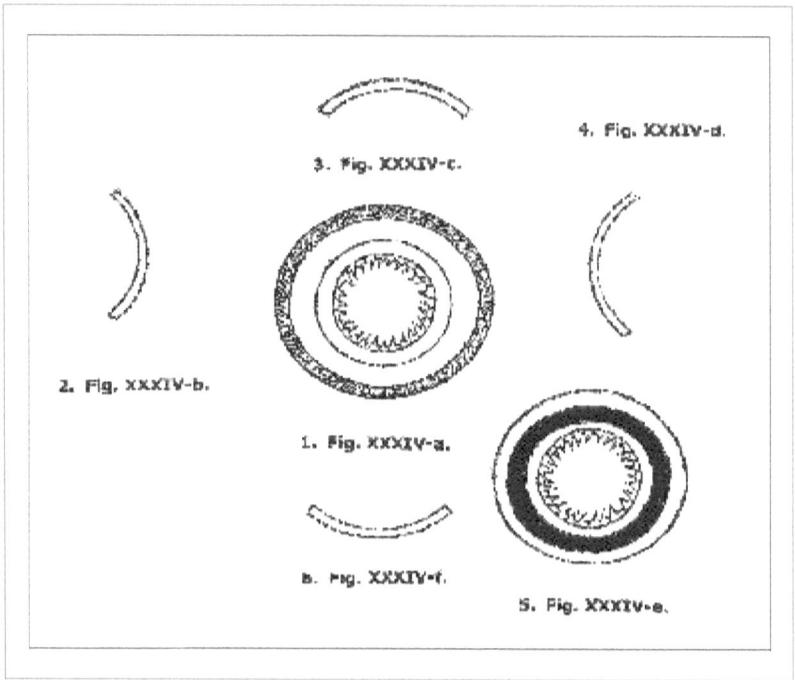

34. FIG. XXXIV.

34. FIG. XXXIV. This figure shown here is of the **Modified Forms Views,** of the **Front of Model X-1** of the fully assembled **Perimeter Saw,** pages **82** to **101**. The parts are described in details on page **88** of the Claims and Specifications.

Page 89. of 140.

Tropical Saws & Export Corporation ™

Perimeter Saw-Design Patent Application
Joe Nathan Brown-Inventor, Designer, Author, Engineering

These are the CLAIMS AND SPECIFICATIONS I-1.

35. FIG. XXXV. This figure shown here is of the **Modified Forms View**, Model-X of the fully assembled, and Perimeter Saw.

1. Fig. XXXV.-a. The Blades' with its' Cutting Edges;

2. Fig. XXXV-b The Blades' Left Side View of the outer Diameter;

3. Fig. XXXV-c The Blades' Stability Circle on the rear View;

4. Fig. XXXV-d The Blades' Top Plan View of the outer diameter;

5. Fig. XXX-e The Blades' Bottom Plan View of the outer diameter;

35.1. The Blades' with its' Cutting Edges, the Blades' full circular View of the outer diameter, shown here details the form of the blade being a circle with a 1/16" to 1/2" inch depth. The Gears on this blade are rectangular in shape.

35.2. The Blades' Left Side View of the outer Diameter, the Blades' Left Side View of the outer diameter, shown here details the form of the blade with a look at a piece with a length of 90 degrees.

35.3. The Blades' Stability Circle on the rear View, this view indicates that the Blade Stability Circle has to be for enough from the cutting edges to leave room for the radius of the pipe, tubing or other material passing through the hole in the center of the blade.

35.4. The Blades' Top Plan View of the outer diameter, The Top Plan View of the Circular Blade, shown here to indicate the circular form of the blade.

35.5. The Blades' Bottom Plan View of the outer diameter, shown here details the form of the blade with a look at a piece with a length of 90 degrees.

Page 90. of 140.

Tropical Saws & Export Corporation ™

Perimeter Saw © ®

Application Drawings and Prints
Joe Nathan Brown-Inventor, Designer, Author, Engineering
These are the CLAIMS AND SPECIFICATIONS I-1.

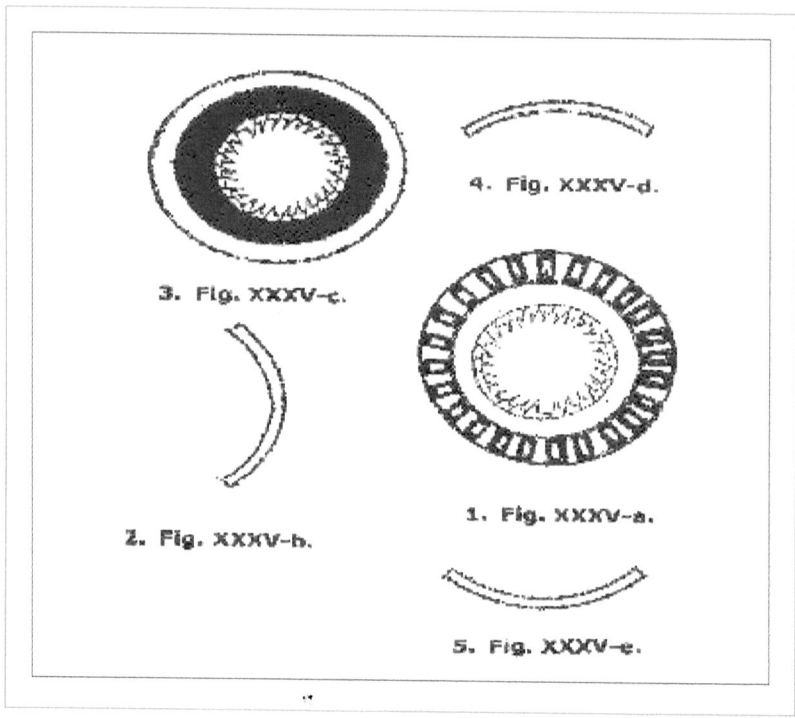

3. Fig. XXXV-c.

4. Fig. XXXV-d.

2. Fig. XXXV-b.

1. Fig. XXXV-a.

5. Fig. XXXV-e.

35. FIG. XXXV.

35. FIG. XXXV This figure shown here is of the **Modified Forms Views,** of the **Front of Model X-1** of the fully assembled **Perimeter Saw,** pages **82** to **101**. The parts are described in details on page **90** of the Claims and Specifications.

Page 91. of 140.

Tropical Saws & Export Corporation ™

Perimeter Saw-Design Patent Application
Joe Nathan Brown-Inventor, Designer, Author, Engineering

These are the CLAIMS AND SPECIFICATIONS I-1.
February 23, 2009

36. FIG. XXXVI. This figure shown here is of the **Modified Forms View**, Model-X of the fully assembled, perimeter Saw.

1. Fig. XXXVI.-a. The Blades' Cutting Edges in a circular diagram for counter clock blade rotational cutting, is the Forward Switching capability.

2. Fig. XXXVI-b The Blades' Cutting Edges in a circular diagram for clockwise blade rotational cutting, is the Reverse Switching capability.

3. Fig. XXXVI-c The Blades' Cutting Edges in a circular diagram for counter clockwise blade rotational cutting, is the Forward Switching capability.

Tropical Saws & Export Corporation ™

Perimeter Saw © ®

Application Drawings and Prints
Joe Nathan Brown-Inventor, Designer, Author, Engineering
These are the CLAIMS AND SPECIFICATIONS I-1.

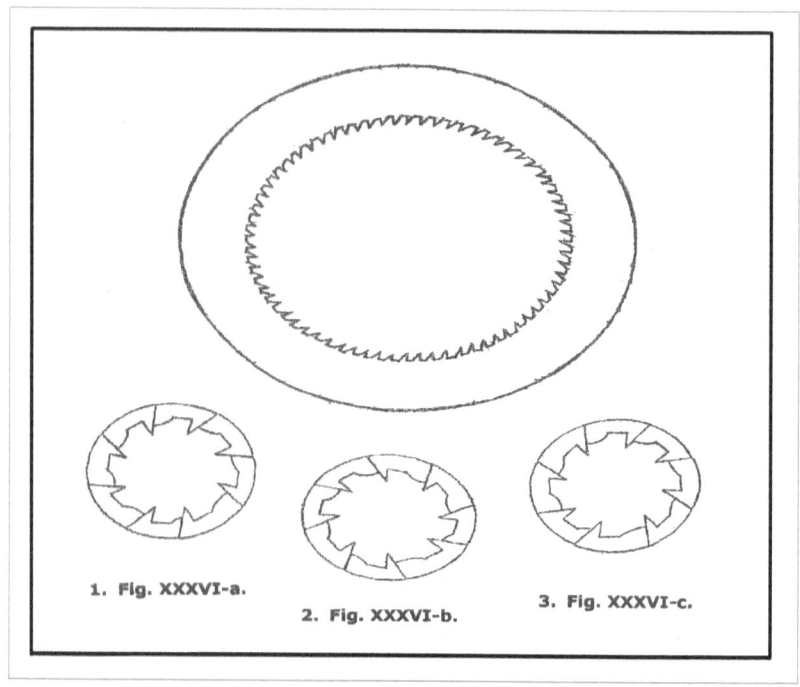

1. Fig. XXXVI-a.

2. Fig. XXXVI-b.

3. Fig. XXXVI-c.

36. FIG. XXXVI.

36. FIG. XXXVI This figure shown here is of the **Modified Forms Views,** of the **Front of Model X-1** of the fully assembled **Perimeter Saw,** pages **82** to **101.** The parts are described in details on page **92** of the Claims and Specifications.

Tropical Saws & Export Corporation ™

Perimeter Saw-Design Patent Application
Joe Nathan Brown-Inventor, Designer, Author, Engineering

These are CLAIMS AND SPECIFICATIONS I-1.

37. FIG. XXXVII. .   This figure shown here is of the **Modified Forms View**, Model-X of the fully assembled, and Perimeter Saw.

1. Fig. XXXVII-a The Blades' Cutting Edges type #1;

2. Fig. XXXVII-b The Blades' Cutting Edges type #2;

3. Fig. XXXVII-c The Blades' Cutting Edges type #3;

4. Fig. XXXVII-d The Blades' Cutting Edges type #4;

5. Fig. XXXVII-e The Blades' Cutting Edges type #5;

37.1. The Blades' Cutting Edges type #1. This type cutting edge will be the standard look in design for the teeth and it is on page 66, 67, 68, 69, 70, 71 92, 93, 94 and 95 of the Drawings, and Claims and Specifications. This type has a circular piece the lifts the discarded materials after the cut.

37.2. The Blades' Cutting Edges type #2. This cutting edge will be the second most used type of the cutting edges because it has a straight piece between the cutting edges.

37.3. The Blades' Cutting Edges type #3. This cutting edges picture shown here makes the fragment of the sharpened cutting edge a triangular shape; which allows for the an easy start in the cutting process.

37.4. The Blades' Cutting Edges type #4. This figure shown here, will allow that a score be made in the material while the blade is rotation to be the cut, with a double edge.

37.5. The Blades' Cutting Edges type #5. This figure shown here, will cause that the blade have another possible to score a mark on the materials it is used on.

Tropical Saws & Export Corporation ™

Perimeter Saw © ®

Application Drawings and Prints
Joe Nathan Brown-Inventor, Designer, Author, Engineering
These are the CLAIMS AND SPECIFICATIONS I-1.

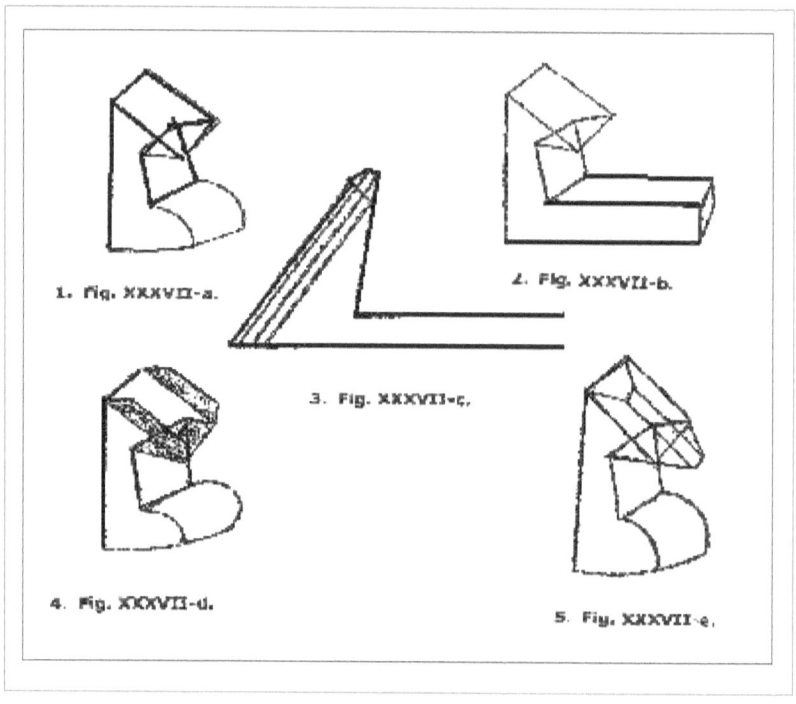

1. Fig. XXXVII-a.

2. Fig. XXXVII-b.

3. Fig. XXXVII-c.

4. Fig. XXXVII-d.

5. Fig. XXXVII-e.

37. FIG. XXXVII.

37. FIG. XXXVII. This figure shown here is of the **Modified Forms Views,** of the **Front of Model X-1** of the fully assembled **Perimeter Saw,** pages **82** to **101**. The parts are described in details on page **94** of the Claims and Specifications.

Tropical Saws & Export Corporation ™

Perimeter Saw-Design Patent Application
Joe Nathan Brown-Inventor, Designer, Author, Engineering

These are the CLAIMS AND SPECIFICATIONS I-1.

38. FIG. XXXVIII. . This figure shown here is of the **Modified Forms View**, Model-X of the fully assembled, perimeter Saw.

1. Fig. XXXVIII.-a is of The Blade Stability Circle;

2. Fig. XXXVIII-b The Components Housing Insert Attachment, Stability Circle;

3. Fig. XXXVIII-c The Friction Absorbing Pad, Screw Mounting Type;

4. Fig. XXXVIII-d The Friction Absorbing Pad, Adhesive Type;

38.1. The Blade Stability Circle seen here separated from the two parts it will be added to gives a good idea that in itself is a separate named piece and not another name for the same piece, Fig. XXXVIII-a

38.2. The Components Housing Insert Attachment, Stability Circle seen here separated from the two parts it will be added to gives some idea that in itself is a separate piece and not another name for the same piece, Fig. XXXVIII-b

38.3. The Friction Absorbing Pad, Screw Mounting Type, can be attached to the Inside Rear Component Insert Attachment, the Blade and the Blade holding Apparatus to perform its design functions and specifications.

38.4. The Friction Absorbing Pad, Adhesive Type, can be attached to the Inside Rear Component Insert Attachment, the Blade and the Blade holding Apparatus to perform it' design functions and specification..

Tropical Saws & Export Corporation ™

Perimeter Saw © ®

Application Drawings and Prints
Joe Nathan Brown-Inventor, Designer, Author, Engineering
These are the CLAIMS AND SPECIFICATIONS I-1.

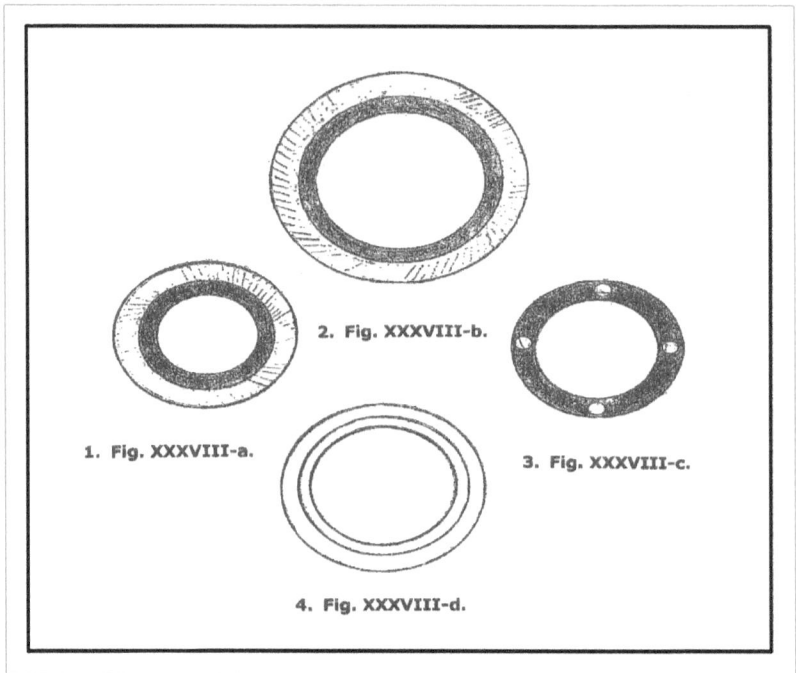

2. Fig. XXXVIII-b.

1. Fig. XXXVIII-a.

3. Fig. XXXVIII-c.

4. Fig. XXXVIII-d.

38. FIG. XXXVIII.

38. FIG. XXXVIII. This figure shown here is of the **Alternate Positional View,** of the **Front of Model X-1** of the fully assembled **Perimeter Saw,** pages **82** to **101**. The parts are described in details on page **96** of the Claims and Specifications.

Tropical Saws & Export Corporation ™

Perimeter Saw-Design Patent Application
Joe Nathan Brown-Inventor, Designer, Author, Engineering

These are the CLAIM AND SPECIFICATION I-1.
February 23, 2009

39. FIG. XXXIX. . This figure shown here is of the **Modified Forms View**, Model-X of the fully assembled, perimeter Saw. It is the Blade Holding Insert Attachment as if it were a separate part. This part is in the top region of the Blade Housing Insert Attachment, molded into the one piece and part of the frame.

This part has the Blade Stability Circle the raise area made to relieve most of

The friction experienced while the blade is rotating.

The Component Housing Insert attachment, has room for the friction pad.

This Absorbing Pad to be molded to it by means of solid molding. It is screw

tighten to the Components Housing Insert Attachment and it is glued or molded

To this section and it will remain in place while the tool is being used.

Tropical Saws & Export Corporation ™

Perimeter Saw © ®

Application Drawings and Prints
Joe Nathan Brown-Inventor, Designer, Author, Engineering
These are the CLAIMS AND SPECIFICATIONS I-1.

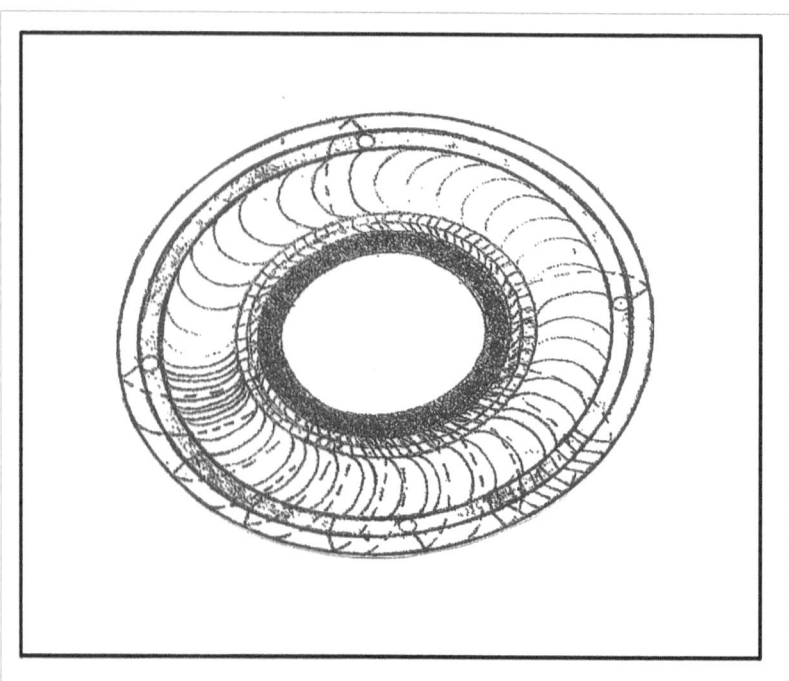

39. FIG. XXXIX.

39. FIG. XXXIX This figure shown here is of the **Modified Forms Views,** of the **Front of Model X-1** of the fully assembled **Perimeter Saw,** pages **82** to **101**. The parts are described in details on page **98** of the Claims and Specifications.

Tropical Saws & Export Corporation ™

Perimeter Saw-Design Patent Application
Joe Nathan Brown-Inventor, Designer, Author, Artist, Engineering

These are the CLAIM AND SPECIFICATION I-1.
May 27, 2009

40. FIG. XL   This figure shown here is of the **Modified Forms View**, Model-X of the fully assembled, perimeter Saw.

1. Fig. XL-a Gear Holes, Diagram;

2. Fig. XL-b Leverage Activators, Diagram;

3. Fig. XL-c Blade Holding Apparatus, Diagram;

4. Fig. XL-d Blade Cutting Edges or Teeth, Diagram;

40.1. The Gear Holes are engineered with square, rectangular and round openings. They will have a depth of: 1/32", 1/16', 1/32 + 1/16, and 1/8" of an inch, depending on the thick of blade stock they are cut from, FIG. XL.

40.2. The gear Leverage Activators are the triangular shaped pieces of the outer circumference. This part touches the gear leverage activators on the elbow gear to turn the blade, FIG. XL.

40.3. The Blade Holding Apparatus, Diagram here is to show where the piece will fit over the top of the blade and where the spacing arrangement will be once this is done, FIG. XL.

40.4. The Blade Cutting Edges or Teeth, Diagram here give a clear picture of the radius surrounding the material placing opening. And it shows the cutting edges using the space remaining after the other parts are installed over the blade to keep it in the saw.

Tropical Saws & Export Corporation ™

Perimeter Saw © ®

Application Drawings and Prints
Joe Nathan Brown-Inventor, Designer, Author, Engineering
These are the CLAIMS AND SPECIFICATIONS I-1.

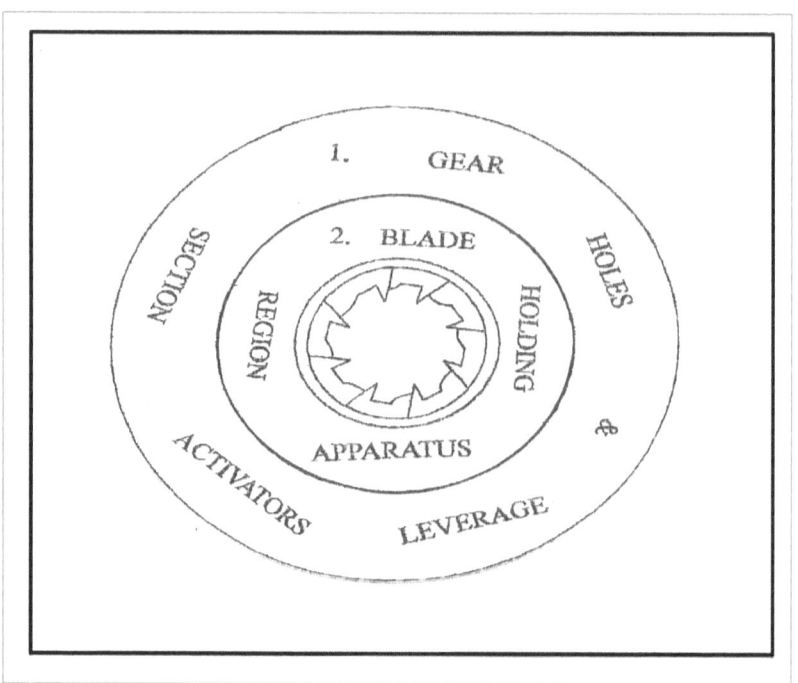

40. FIG. XL.

40. FIG. XL This figure shown here is of the **Modified Forms Views,** of the **Front of Model X-1** of the fully assembled **Perimeter Saw,** pages **82** to **101.** The parts are described in details on page **100** of the Claims and Specifications.

Tropical Saws & Export Corporation ™

Perimeter Saw-Design Patent Application
Joe Nathan Brown-Inventor, Designer, Author, Artist, Engineering

The Specifications of the **Lengths,** 12 Sizes and Classifications
of the Perimeter Saw is below.

The **Blades**' outer Diameter; Is listed below.	The **Length** of Saw From Top To Bottom; is Listed below.	
1.	4″	Blade's OD	12″
2.	8″	Blade's OD	14″
3.	12″	Blade's OD	16″
4.	16″	Blade's OD	20″
5.	20″	Blade's OD	26″
6.	24″	Blade's OD	32″
7.	28″	Blade's OD	36″
8.	32″	Blade's OD	40″
9.	36″	Blade's OD	46″
10.	40″	Blade's OD	48″
11.	44″	Blade's OD	48″
12.	48″	Blade's OD	50″

Tropical Saws & Export Corporation ™

Perimeter Saw-Design Patent Application
Joe Nathan Brown-Inventor, Designer, Author, Artist, Engineering

These are the CLAIM AND SPECIFICATION I-1.

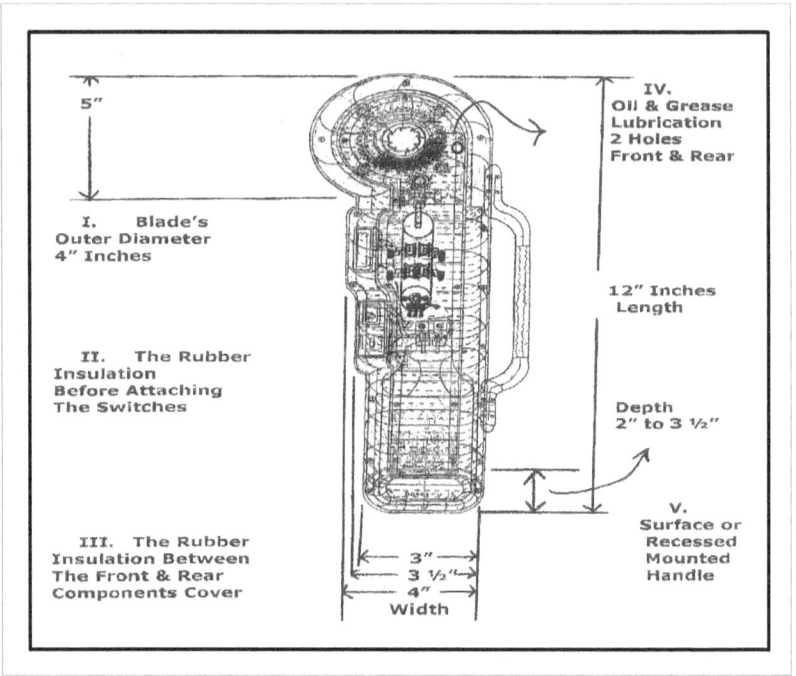

41. FIG. XLI.

41. FIG. XLI Are the specifications of the **Lengths, Widths,** and **Depths** of the **12** Sizes and Classifications of the **Perimeter Saw.** The parts are described in details on page **102** of the claims and specifications the **Lengths** of the **Perimeter Saw.**

Page 103. of 140.

Tropical Saws & Export Corporation ™

Perimeter Saw-Design Patent Application
Joe Nathan Brown-Inventor, Designer, Artist, Author

The **Specifications** of the **Lengths, Widths** & **Depths** of the **12** Sizes and Classifications of the Perimeter Saw. The **Widths** in Inches from the left Side to the Right Side; which is consistent having the handle on the right as the further most part of measurement.

Stage 1	Stage 1	Stage 1	Stage 1	Stage 1
A. The Stage 1 are the Outer Diameters of the blades.				
B. The Stage 2 is the Upper Blade Section Surrounding the Blade.				
C. The Stage 3 is the Push Button Switch section.				
D. The Stage 4 is the Forward/Reversing & Locking Switches section.				
E. The Stage 5 is the Lower Battery section.				

1.	4″	5″	4″	3 ½	3″
2.	8″	9″	8″	4 ½″	4″
3.	12″	13 ½″	12″	4 ½″	4″
4.	16″	17 ½″	16″	4 ½″	4″
5.	20″	22″	20″	5 ½″	4 ½″
6.	24″	26″	24″	5 ½″	4 ½″
7.	28″	30 ½″	28″	5 ½″	5″
8.	32″	34 ½″	32″	5 ½″	5″
9.	36″	38 ½″	36″	6 ½ ″	5″
10.	40″	42 ½″	40″	6 ½″	6″
11.	44″	46 ½″	44″	7 ½″	7″
12.	48″	52″	48″	11 ½ ″	11″

Perimeter Saw-Design Patent Application
Joe Nathan Brown-Inventor, Designer, Author, Artist, Engineering

These are the CLAIM AND SPECIFICATION I-1.

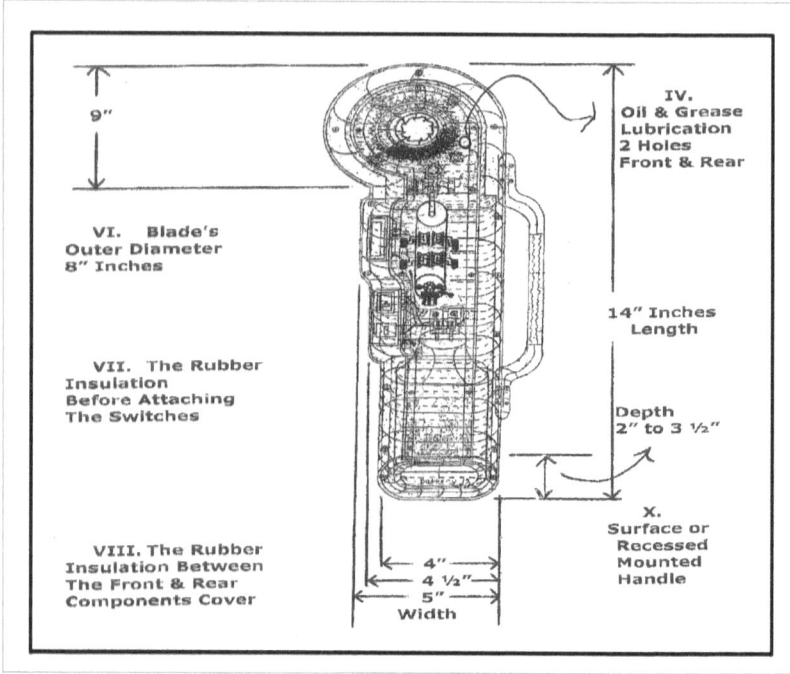

42. FIG. XLII.

42. FIG. XLII Are the specifications of the **Lengths, Widths,** and **Depths** of the **12** Sizes and Classifications of the **Perimeter Saw.** The parts are described in details on page **104** of the claims and specifications, the **Widths** of the **Perimeter Saw.**

Page 105. of 140.

Tropical Saws & Export Corporation ™

Perimeter Saw © ®

Design Patent Applications
Joe Nathan Brown-Inventor, Designer, Artist, Author

The Depth in Inches, From the Front of the Perimeter Saw to the Rear
of the Perimeter Saw; which is consistent throughout.

These are the **Blade's Outer** Diameters.		These are the **Blade's Depth** in Inches.
1.	4″	2″ to 3 1/2″
2.	8″	2″ to 3 1/2″
3.	12″	3″ to 4 1/2″
4.	16″	3″ to 4 1/2″
5.	20″	4″ to 5 1/2″
6.	24″	4″ to 5 1/2″
7.	28″	4 1/2″ to 5 1/2″
8.	32″	4 1/2″ to 5 1/2″
9.	36″	4 1/2″ to 5 1/2″
10.	40″	4 1/2″ to 5 1/2″
11.	44″	4 1/2″ to 6″
12.	48″	4 1/2″ to 6″

Tropical Saws & Export Corporation ™

Perimeter Saw-Design Patent Application
Joe Nathan Brown-Inventor, Designer, Author, Artist, Engineering

These are the CLAIM AND SPECIFICATION I-1.

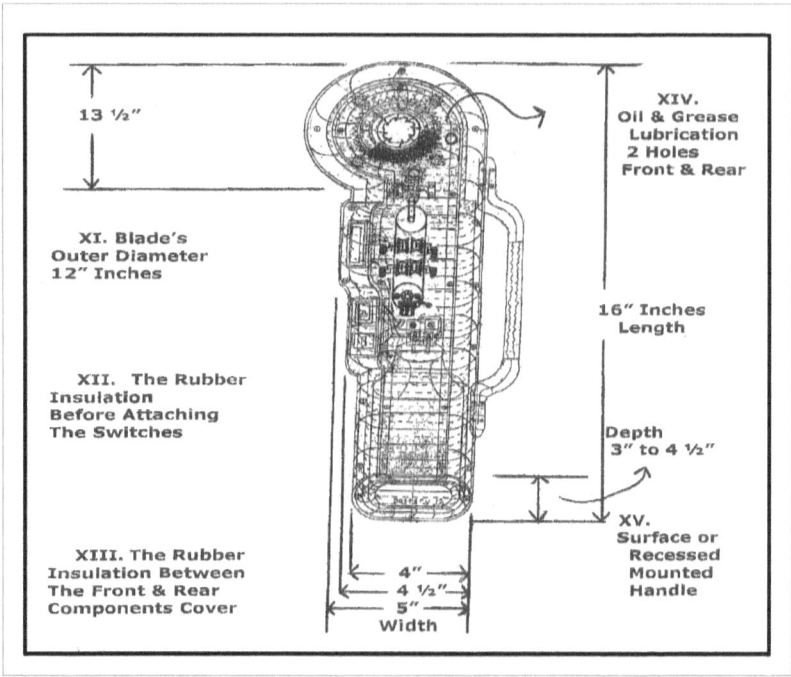

43. FIG. XLIII.

43. FIG. XLIII Are the specifications of the **Lengths, Widths,** and **Depths** of the **12** Sizes and Classifications of the **Perimeter Saw.** The parts are described in details on page **106** of the claims and specifications, the **Depths** of the **Perimeter Saw.**

Tropical Saws & Export Corporation ™

Perimeter Saw-Design Patent Application
Joe Nathan Brown-Inventor, Designer, Author, Artist, Engineering

The Specifications of the **Lengths,** 12 Sizes and Classifications
of the Perimeter Saw is below.

The **Blades**' outer Diameter; Is listed below.		The **Length** of Saw From Top To Bottom; is Listed below.
1.	4″ Blade's OD	12″
2.	8″ Blade's OD	14″
3.	12″ Blade's OD	16″
4.	16″ Blade's OD	20″
5.	20″ Blade's OD	26″
6.	24″ Blade's OD	32″
7.	28″ Blade's OD	36″
8.	32″ Blade's OD	40″
9.	36″ Blade's OD	46″
10.	40″ Blade's OD	48″
11.	44″ Blade's OD	48″
12.	48″ Blade's OD	50″

Tropical Saws & Export Corporation ™

Perimeter Saw-Design Patent Application

Joe Nathan Brown-Inventor, Designer, Author, Artist, Engineering

These are the CLAIM AND SPECIFICATION I-1.

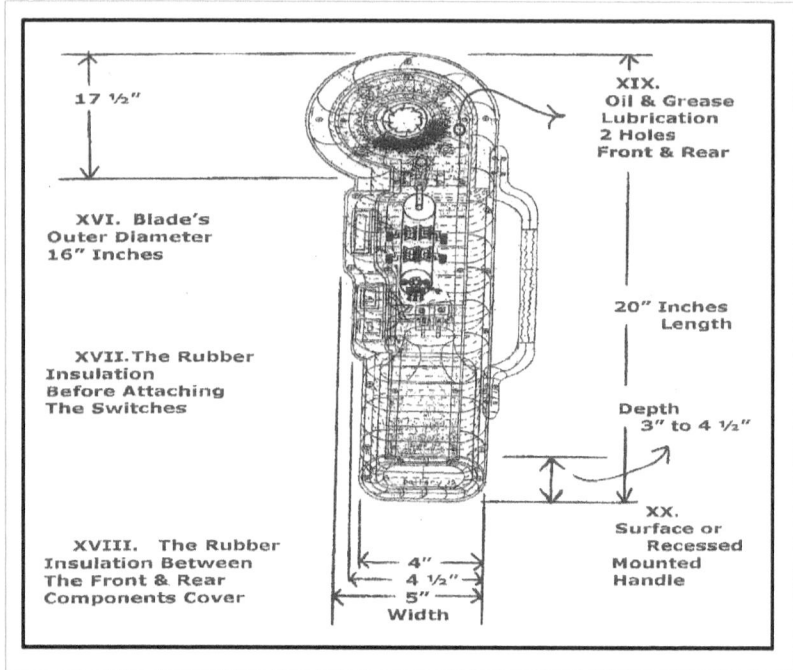

17 ½"

XVI. Blade's
Outer Diameter
16" Inches

XVII.The Rubber
Insulation
Before Attaching
The Switches

XVIII.   The Rubber
Insulation Between
The Front & Rear
Components Cover

XIX.
Oil & Grease
Lubrication
2 Holes
Front & Rear

20" Inches
Length

Depth
3" to 4 ½"

XX.
Surface or
Recessed
Mounted
Handle

4"
4 ½"
5"
Width

44. FIG. XLIV.

44. FIG. XLIV Are the specifications of the **Lengths, Widths,** and **Depths** of the **12** Sizes and Classifications of the **Perimeter Saw.**   The parts are described in details on page **108** of the claims and specifications, the **Lengths** of the **Perimeter Saw.**

Page 109. of 140.

Tropical Saws & Export Corporation ™

Perimeter Saw-Design Patent Application
Joe Nathan Brown-Inventor, Designer, Artist, Author

The **Specifications** of the **Lengths, Widths** & **Depths** of the **12** Sizes and
Classifications of the Perimeter Saw. The **Widths** in Inches from the left
Side to the Right Side; which is consistent having the handle on the right
as the further most part of measurement.

Stage 1	Stage 1	Stage 1	Stage 1	Stage 1
A. The Stage 1 are the Outer Diameters of the blades.				
B. The Stage 2 is the Upper Blade Section Surrounding the Blade.				
C. The Stage 3 is the Push Button Switch section.				
D. The Stage 4 is the Forward/Reversing & Locking Switches section.				
E. The Stage 5 is the Lower Battery section.				
1. 4"	5"	4"	3 ½	3"
2. 8"	9"	8"	4 ½"	4"
3. 12"	13 ½"	12"	4 ½"	4"
4. 16"	17 ½"	16"	4 ½"	4"
5. 20"	22"	20"	5 ½"	4 ½"
6. 24"	26"	24"	5 ½"	4 ½"
7. 28"	30 ½"	28"	5 ½"	5"
8. 32"	34 ½"	32"	5 ½"	5"
9. 36"	38 ½"	36"	6 ½ "	5"
10. 40"	42 ½"	40"	6 ½"	6"
11. 44"	46 ½"	44"	7 ½"	7"
12. 48"	52"	48"	11 ½ "	11"

Page 110. of 140.

Tropical Saws & Export Corporation ™

Perimeter Saw-Design Patent Application
Joe Nathan Brown-Inventor, Designer, Author, Artist, Engineering

These are the CLAIM AND SPECIFICATION I-1.

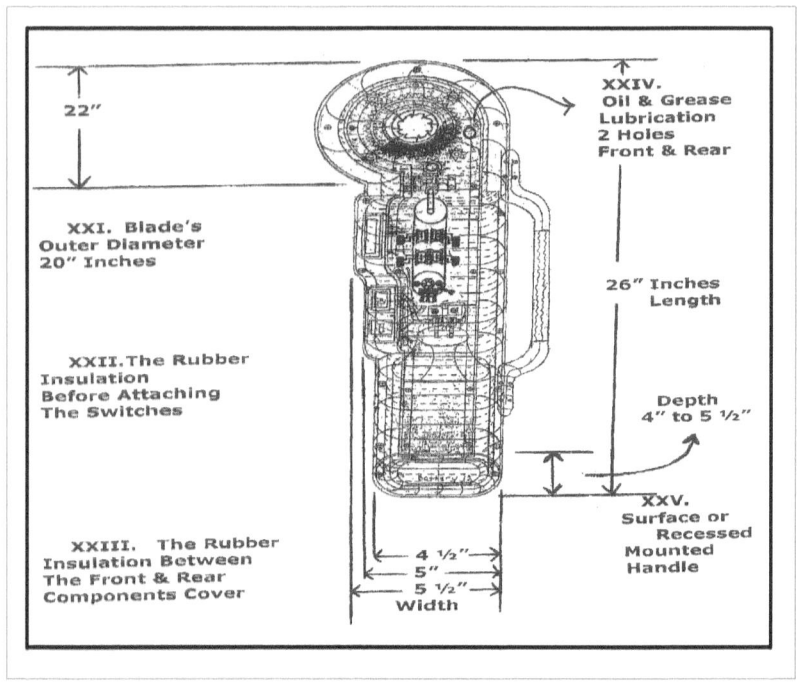

45. FIG. XLV.

45. FIG. XLV Are the specifications of the **Lengths, Widths,** and **Depths** of the **12** Sizes and Classifications of the **Perimeter Saw.** The parts are described in details on page **110** of the claims and specifications, the **Widths** of the **Perimeter Saw.**

Page 111. of 140.

Tropical Saws & Export Corporation ™

Perimeter Saw © ®

Design Patent Applications
Joe Nathan Brown-Inventor, Designer, Artist, Author

The Depth in Inches, From the Front of the Perimeter Saw to the Rear
of the Perimeter Saw; which is consistent throughout.

These are the **Blade's Outer** Diameters.	These are the **Blade's Depth** in Inches.
1. 4"	2" to 3 1/2"
2. 8"	2" to 3 1/2"
3. 12"	3" to 4 1/2"
4. 16"	3" to 4 1/2"
5. 20"	4" to 5 1/2"
6. 24"	4" to 5 1/2"
7. 28"	4 1/2" to 5 1/2"
8. 32"	4 1/2" to 5 1/2"
9. 36"	4 1/2" to 5 1/2"
10. 40"	4 1/2" to 5 1/2"
11. 44"	4 1/2" to 6"
12. 48"	4 1/2" to 6"

Tropical Saws & Export Corporation ™

Perimeter Saw-Design Patent Application
Joe Nathan Brown-Inventor, Designer, Author, Artist, Engineering

These are the CLAIM AND SPECIFICATION I-1.

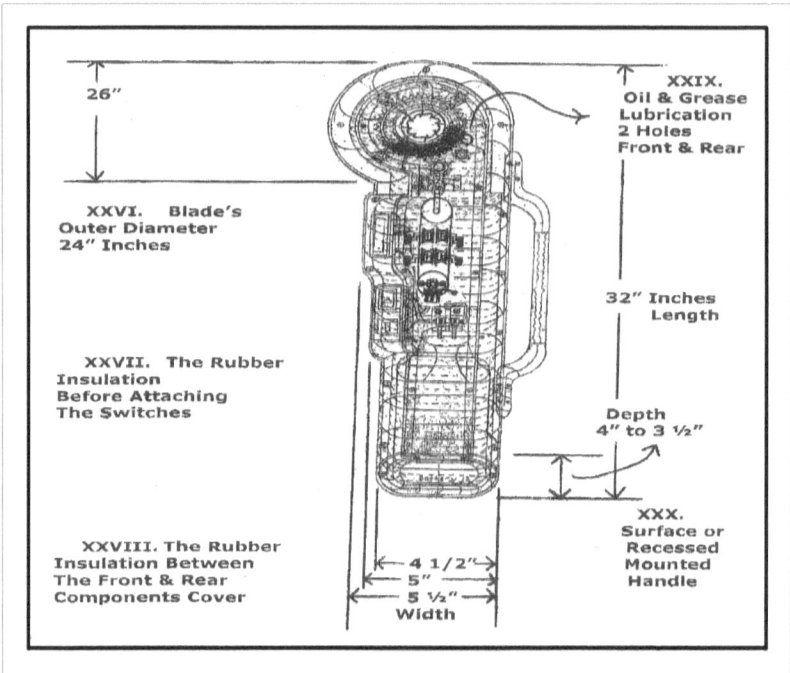

46. FIG. XLVI.

46. FIG. XLVI Are the specifications of the **Lengths, Widths,** and **Depths** of the **12** Sizes and Classifications of the **Perimeter Saw.** The parts are described in details on page **112** of the claims and specifications, the **Depths** of the **Perimeter Saw.**

Tropical Saws & Export Corporation ™

Perimeter Saw-Design Patent Application
Joe Nathan Brown-Inventor, Designer, Author, Artist, Engineering

The Specifications of the **Lengths,** 12 Sizes and Classifications
of the Perimeter Saw is below.

The **Blades'** outer Diameter; Is listed below.		The **Length** of Saw From Top To Bottom; is Listed below.
1.	4″ Blade's OD	12″
2.	8″ Blade's OD	14″
3.	12″ Blade's OD	16″
4.	16″ Blade's OD	20″
5.	20″ Blade's OD	26″
6.	24″ Blade's OD	32″
7.	28″ Blade's OD	36″
8.	32″ Blade's OD	40″
9.	36″ Blade's OD	46″
10.	40″ Blade's OD	48″
11.	44″ Blade's OD	48″
12.	48″ Blade's OD	50″

Tropical Saws & Export Corporation ™

Perimeter Saw-Design Patent Application
Joe Nathan Brown-Inventor, Designer, Author, Artist, Engineering

These are the CLAIM AND SPECIFICATION I-1.

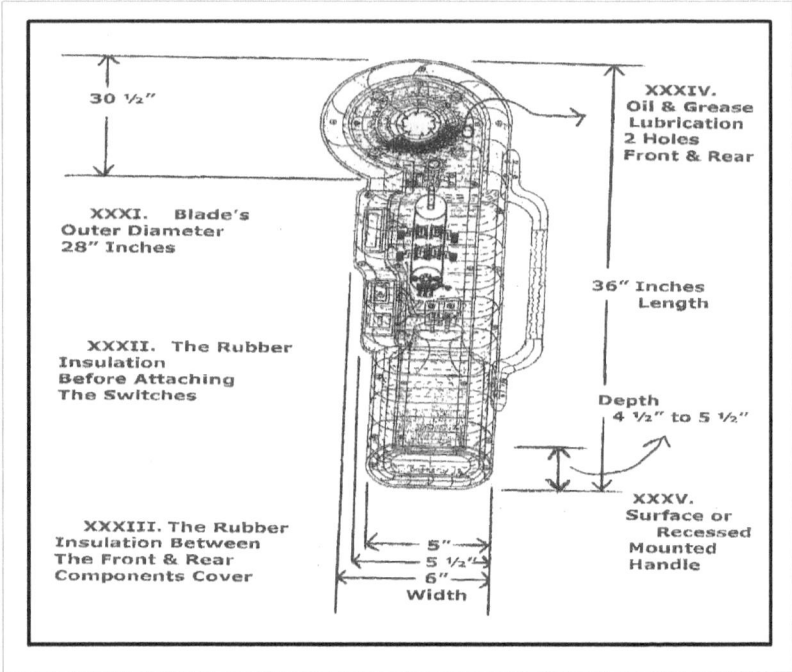

47. FIG. XLVII.

47. FIG. XLVII Are the specifications of the **Lengths, Widths,** and **Depths** of the **12** Sizes and Classifications of the **Perimeter Saw.** The parts are described in details on page **114** of the claims and specifications, the **Lengths** of the **Perimeter Saw.**

Tropical Saws & Export Corporation ™

Perimeter Saw-Design Patent Application
Joe Nathan Brown-Inventor, Designer, Artist, Author

The **Specifications** of the **Lengths, Widths** & **Depths** of the **12** Sizes and Classifications of the Perimeter Saw. The **Widths** in Inches from the left Side to the Right Side; which is consistent having the handle on the right as the further most part of measurement.

Stage 1	Stage 1	Stage 1	Stage 1	Stage 1
A. The Stage 1 are the Outer Diameters of the blades.				
B. The Stage 2 is the Upper Blade Section Surrounding the Blade.				
C. The Stage 3 is the Push Button Switch section.				
D. The Stage 4 is the Forward/Reversing & Locking Switches section.				
E. The Stage 5 is the Lower Battery section.				
1. 4″	5″	4″	3 ½	3″
2. 8″	9″	8″	4 ½″	4″
3. 12″	13 ½″	12″	4 ½″	4″
4. 16″	17 ½″	16″	4 ½″	4″
5. 20″	22″	20″	5 ½″	4 ½″
6. 24″	26″	24″	5 ½″	4 ½″
7. 28″	30 ½″	28″	5 ½″	5″
8. 32″	34 ½″	32″	5 ½″	5″
9. 36″	38 ½″	36″	6 ½ ″	5″
10. 40″	42 ½″	40″	6 ½″	6″
11. 44″	46 ½″	44″	7 ½″	7″
12. 48″	52″	48″	11 ½ ″	11″

Tropical Saws & Export Corporation ™

Perimeter Saw-Design Patent Application
Joe Nathan Brown-Inventor, Designer, Author, Artist, Engineering

These are the CLAIM AND SPECIFICATION I-1.

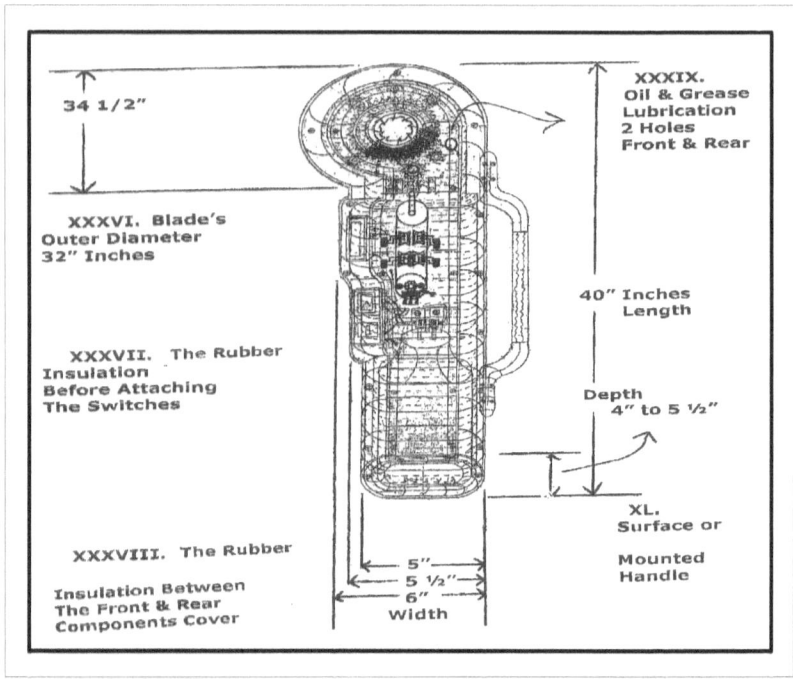

34 1/2"

XXXIX.
Oil & Grease
Lubrication
2 Holes
Front & Rear

XXXVI. Blade's
Outer Diameter
32" Inches

40" Inches
Length

XXXVII. The Rubber
Insulation
Before Attaching
The Switches

Depth
4" to 5 ½"

XL.
Surface or
Mounted
Handle

XXXVIII. The Rubber

5"
5 ½"
6"
Width

Insulation Between
The Front & Rear
Components Cover

48. FIG. XLVIII.

48. FIG. XLVIII Are the specifications of the **Lengths, Widths,** and **Depths** of the **12** Sizes and Classifications of the **Perimeter Saw.** The parts are described in details on page **116** of the claims and specifications, the **Widths** of the **Perimeter Saw.**

Page 117. of 140.

Tropical Saws & Export Corporation ™

Perimeter Saw © ®

Design Patent Applications
Joe Nathan Brown-Inventor, Designer, Artist, Author

The Depth in Inches, From the Front of the Perimeter Saw to the Rear of the Perimeter Saw; which is consistent throughout.

These are the Blade's Outer Diameters.	These are the **Blade's** Depth in Inches.
1. 4″	2″ to 3 1/2″
2. 8″	2″ to 3 1/2″
3. 12″	3″ to 4 1/2″
4. 16″	3″ to 4 1/2″
5. 20″	4″ to 5 1/2″
6. 24″	4″ to 5 1/2″
7. 28″	4 1/2″ to 5 1/2″
8. 32″	4 1/2″ to 5 1/2″
9. 36″	4 1/2″ to 5 1/2″
10. 40″	4 1/2″ to 5 1/2″
11. 44″	4 1/2″ to 6″
12. 48″	4 1/2″ to 6″

Tropical Saws & Export Corporation ™

Perimeter Saw-Design Patent Application
Joe Nathan Brown-Inventor, Designer, Author, Artist, Engineering

These are the CLAIM AND SPECIFICATION I-1.

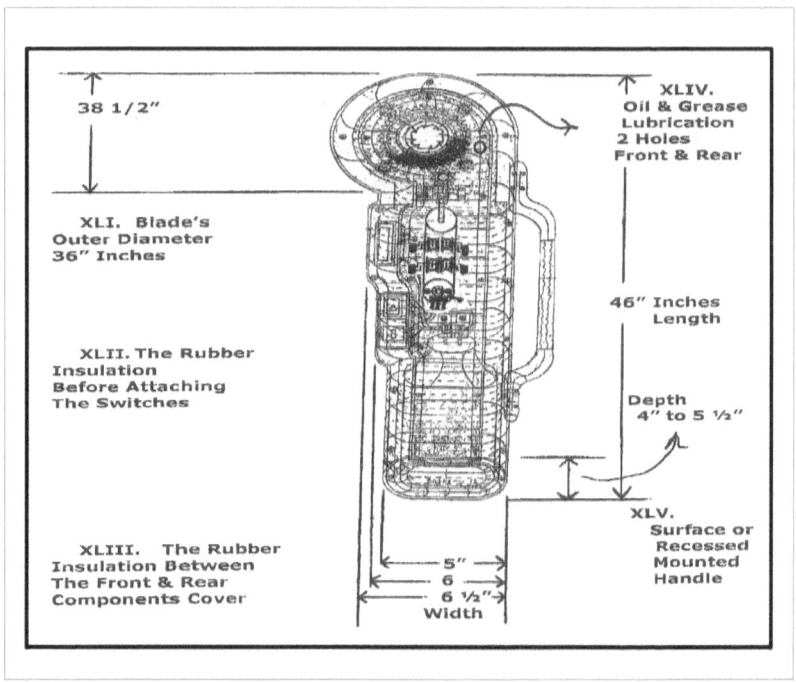

49. FIG. XLIX.

49. FIG. XLIX Are the specifications of the **Lengths, Widths,** and **Depths** of the **12** Sizes and Classifications of the **Perimeter Saw.** The parts are described in details on page **118** of the claims and specifications, the **Depths** of the **Perimeter Saw.**

Page 119. of 140.

Tropical Saws & Export Corporation ™

Perimeter Saw-Design Patent Application
Joe Nathan Brown-Inventor, Designer, Author, Artist, Engineering

The Specifications of the **Lengths,** 12 Sizes and Classifications
of the Perimeter Saw is below.

The **Blades**' outer Diameter; Is listed below.	The **Length** of Saw From Top To Bottom; is Listed below.
1. 4″ Blade's OD	12″
2. 8″ Blade's OD	14″
3. 12″ Blade's OD	16″
4. 16″ Blade's OD	20″
5. 20″ Blade's OD	26″
6. 24″ Blade's OD	32″
7. 28″ Blade's OD	36″
8. 32″ Blade's OD	40″
9. 36″ Blade's OD	46″
10. 40″ Blade's OD	48″
11. 44″ Blade's OD	48″
12. 48″ Blade's OD	50″

Tropical Saws & Export Corporation ™

Perimeter Saw-Design Patent Application
Joe Nathan Brown-Inventor, Designer, Author, Artist, Engineering

These are the CLAIM AND SPECIFICATION I-1.

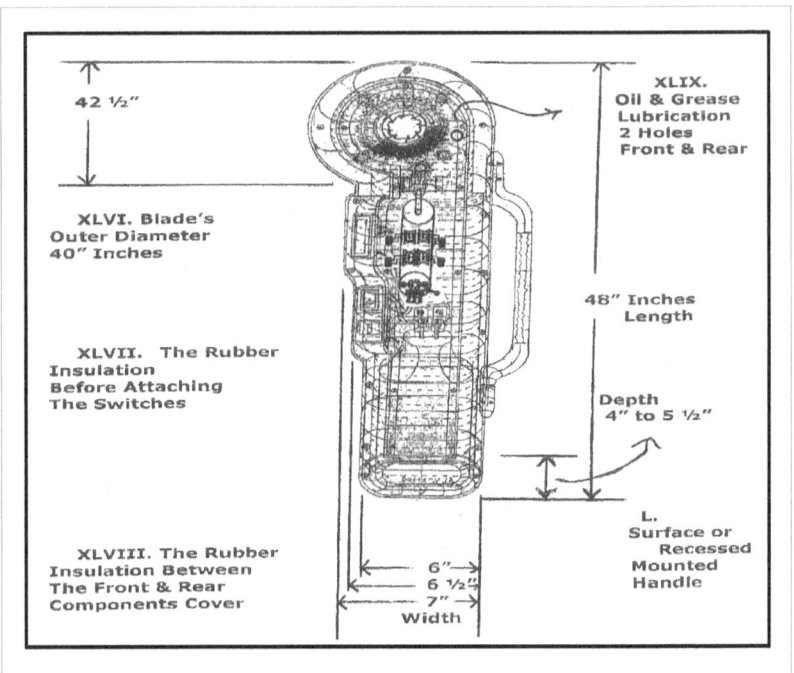

42 ½"

XLIX.
Oil & Grease
Lubrication
2 Holes
Front & Rear

XLVI. Blade's
Outer Diameter
40" Inches

48" Inches
Length

XLVII. The Rubber
Insulation
Before Attaching
The Switches

Depth
4" to 5 ½"

XLVIII. The Rubber
Insulation Between
The Front & Rear
Components Cover

6"
6 ½"
7"
Width

L.
Surface or
Recessed
Mounted
Handle

50. FIG. L.

50. FIG. L Are the specifications of the **Lengths, Widths,** and **Depths** of the **12** Sizes and Classifications of the **Perimeter Saw.** The parts are described in details on page **120** of the claims and specifications, the **Lengths** of the **Perimeter Saw.**

Tropical Saws & Export Corporation ™

Perimeter Saw-Design Patent Application
Joe Nathan Brown-Inventor, Designer, Artist, Author

The **Specifications** of the **Lengths, Widths** & **Depths** of the **12** Sizes and Classifications of the Perimeter Saw. The **Widths** in Inches from the left Side to the Right Side; which is consistent having the handle on the right as the further most part of measurement.

Stage 1	Stage 1	Stage 1	Stage 1	Stage 1
A. The Stage 1 are the Outer Diameters of the blades.				
B. The Stage 2 is the Upper Blade Section Surrounding the Blade.				
C. The Stage 3 is the Push Button Switch section.				
D. The Stage 4 is the Forward/Reversing & Locking Switches section.				
E. The Stage 5 is the Lower Battery section.				
1. 4″	5″	4″	3 ½	3″
2. 8″	9″	8″	4 ½″	4″
3. 12″	13 ½″	12″	4 ½″	4″
4. 16″	17 ½″	16″	4 ½″	4″
5. 20″	22″	20″	5 ½″	4 ½″
6. 24″	26″	24″	5 ½″	4 ½″
7. 28″	30 ½″	28″	5 ½″	5″
8. 32″	34 ½″	32″	5 ½″	5″
9. 36″	38 ½″	36″	6 ½ ″	5″
10. 40″	42 ½″	40″	6 ½″	6″
11. 44″	46 ½″	44″	7 ½″	7″
12. 48″	52″	48″	11 ½ ″	11″

Perimeter Saw-Design Patent Application

Joe Nathan Brown-Inventor, Designer, Author, Artist, Engineering

These are the CLAIM AND SPECIFICATION I-1.

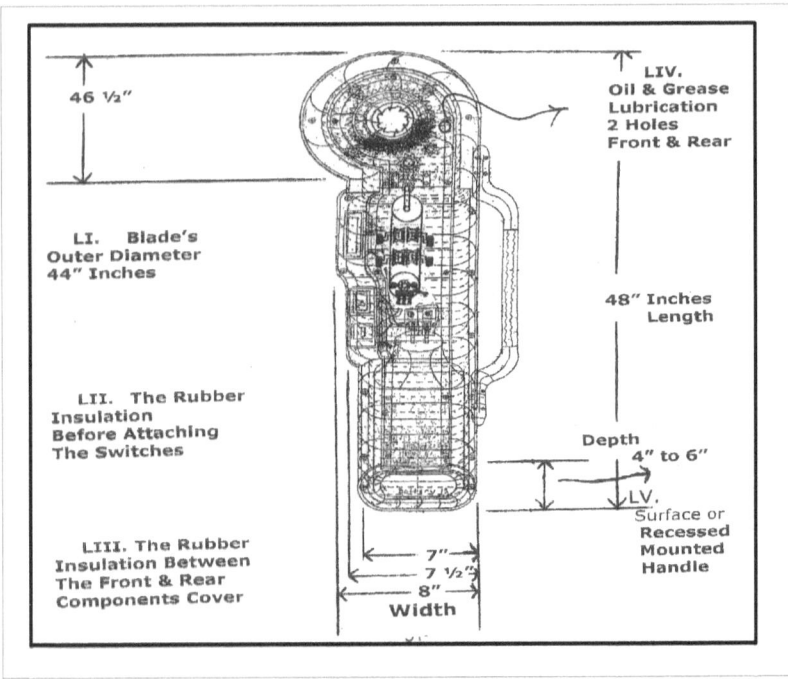

46 ½"

LIV.
Oil & Grease
Lubrication
2 Holes
Front & Rear

LI. Blade's
Outer Diameter
44" Inches

48" Inches
Length

LII. The Rubber
Insulation
Before Attaching
The Switches

Depth
4" to 6"

LV.
Surface or
Recessed
Mounted
Handle

LIII. The Rubber
Insulation Between
The Front & Rear
Components Cover

7"
7 ½"
8"
Width

51. FIG. LI.

51. FIG. LI. Are the specifications of the **Lengths, Widths,** and **Depths** of the **12** Sizes and Classifications of the **Perimeter Saw.** The parts are described in details on page **122** of the claims and specifications, the **Widths** of the **Perimeter Saw.**

Tropical Saws & Export Corporation ™

Perimeter Saw © ®

Design Patent Applications
Joe Nathan Brown-Inventor, Designer, Artist, Author

The Depth in Inches, From the Front of the Perimeter Saw to the Rear of the Perimeter Saw; which is consistent throughout.

These are the **Blade's Outer** Diameters.	These are the **Blade's Depth** in Inches.
1. 4″	2″ to 3 1/2″
2. 8″	2″ to 3 1/2″
3. 12″	3″ to 4 1/2″
4. 16″	3″ to 4 1/2″
5. 20″	4″ to 5 1/2″
6. 24″	4″ to 5 1/2″
7. 28″	4 1/2″ to 5 1/2″
8. 32″	4 1/2″ to 5 1/2″
9. 36″	4 1/2″ to 5 1/2″
10. 40″	4 1/2″ to 5 1/2″
11. 44″	4 1/2″ to 6″
12. 48″	4 1/2″ to 6″

Tropical Saws & Export Corporation ™

Perimeter **S**aw-**D**esign **P**atent **A**pplication
Joe Nathan Brown-Inventor, Designer, Author, Artist, Engineering

These are the CLAIM AND SPECIFICATION I-1.

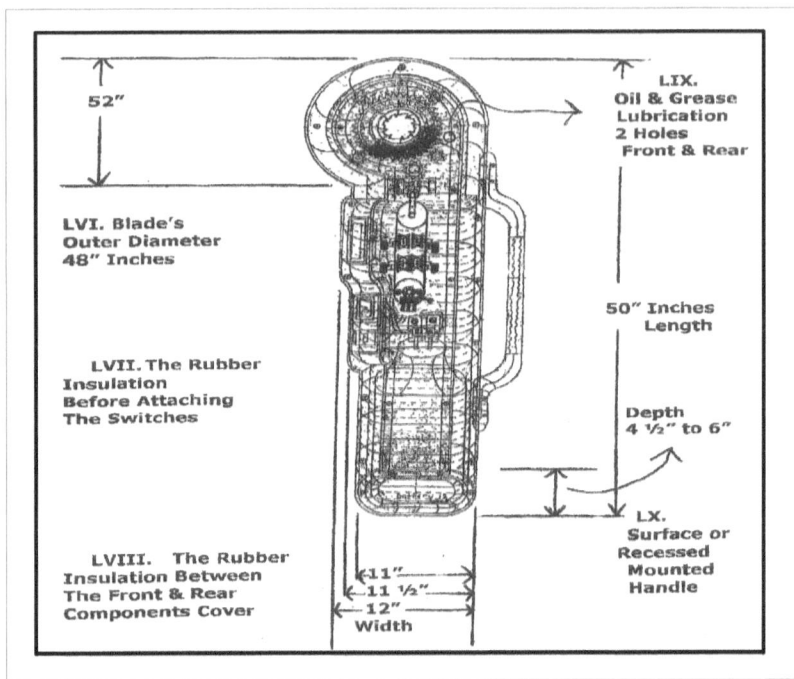

52. FIG. LII.

52. FIG. LII. Are the specifications of the **Lengths, Widths,** and **Depths** of the **12** Sizes and Classifications of the **Perimeter Saw.** The parts are described in details on page **124** of the claims and specifications, the **Depths** of the **Perimeter Saw.**

Tropical Saws & Export Corporation ™

Perimeter Saw-Design Patent Application
Joe Nathan Brown-Inventor, Designer, Author, Engineering

These are the CLAIMS AND SPECIFICATION I-1.

PARTS LISTINGS-and Descriptions

7. Fig. VII is from page 29 of the claims I-1. & specifications

7.1-1 THE BLADE STABILITY GEARS-There are four 4 of these gears that are placed evenly surrounding the blade to keep it in perfect alignment while it is rotating. And oil could be added in this area when a where need to keep the parts lubricated.

7.1-2. THE BLADE WITH CUTTING EDGES, STABILITY CIRCLE & FRICTION PAD- The Blade has as many cutting edges as there are gears. Starting with 12 cutting edges up to 100 of them is a full blade. The cutting edges will be sharpened on the flat surface and also on a triangle reduction. This will allow for them to cut smaller and finer material with an easy start.

7.2-2 THE BLADE STABILITY CIRCLE- The Blade Stability Circle is the part of the blade the keeps the blade rotating in perfect symmetry. It is on the rear of the blade and it is an important piece of it. It is made with this raised platform because without it the blade would be turned by the gears, and would be less stable for cutting materials. As the heavy pressure from the gears will make the blade have an unbalance rotation.

Tropical Saws & Export Corporation ™

Perimeter Saw-Design Patent Application
Joe Nathan Brown-Inventor, Designer, Author, Engineering

These are the CLAIMS AND SPECIFICATIONS I-1.

PARTS LISTINGS-and Descriptions

7. This is Fig. VII from page 29, of the Claims I-1. & Specifications

7.2-2. BLADE STABILITY CIRCLE- The Blade Stability Circle is located on the rear of the blade. It is a circle conforming to the symmetric precision of the part this piece is attached to. It can be metal tool cut into the blade where as a 1/16" of an inch is removed form a 1/8", 1/8" + 1/16" or 1/4" thickness blade in a circular form. The Blade Stability Circle must match the Component Housing Insert Attachment, Stability Circle and it will fit in the inner portion of this part. So this will cause that pressure from the Blade Stability Circle while it is rotating with the Blade not to exceed the boundary of the Components Housing Insert Attachment, Stability.

7.2-2-a is the stability Circle on the rear of the blade is raised from off the back of the blade 1/32", 1/16", 1/16" + 1/32" and 1/8" of an inch. This allows for enough of the blade's metal to be placed in the circle to form a good rotational guide, for the Stability Circle in the Blade Holding Apparatus a part of the Frame.

7.2-2-b The Blade Friction Absorbing Pad, is a piece that is screw attached to the Blade or bonded to it; which reduces the friction of the metal to metal contact while the blade is rotating in the Blade Holding Apparatus in the Top Frame Section.

Tropical Saws & Export Corporation ™

Perimeter Saw-Design Patent Application
Joe Nathan Brown-Inventor, Designer, Author, Engineering

These are the CLAIMS AND SPECIFICATIONS I-1.
7. This is Fig. VII from page 29, of the Claims I-1. & Specifications

PARTS LISTINGS-and Descriptions

7.1-3. THE REAR COMPONENTS INSERT ATTACHMENT & FRAME- Is the part that the Blade, Blade Stability Circle, Blade Stability Gears, Friction Absorbing Pad, 90 Degree Elbow Gear, AC, DC Motor, Push Button Switch, Forward/Reverse Switch, Position Locking Switch, 12volt, 14volt, 18volt, 24volt, 48volt Battery, Battery Plug In Terminals, 120 volt Cord Plug In Terminals and Component Bottom Cover are attached to.

7.1-4. THE BLADE STABILITY GEARS-There are four 4 of these gears that are placed evenly surrounding the blade to keep it in perfect alignment while it is rotating. And oil could be added in this area when a where need to keep the part lubricated.

7.1-5. THE PUSH BUTTON ON/OFF SWITCH-Is located on the left side of the fully assembled perimeter saw. And it will be used when a job has a need for the start and stopping of the motor by the finger pressure of the thumb or any finger nearest to it.

7.1-6. is of the FORWARD & REVERSE SWITCH-This switch is made and designed to cause the motor to rotate the blade in two directions. This will allow the capability of two blade types 1 cutting in clockwise direction and the other cutting in a counter-clockwise direction.

Tropical Saws & Export Corporation ™

Perimeter Saw-Design Patent Application
Joe Nathan Brown-Inventor, Designer, Author, Engineering

These are the CLAIMS AND SPECIFICATIONS I-1.
February 23, 2009

7. This is Fig. VII from page 29, of the Claims I-1. & Specifications

PARTS LISTINGS-and Descriptions

7.1-7 POSITION LOCKING ON/OFF SWITCH- the Position Locking Switch is design to allow the switch to be put in an on position to make the saw motor turn. It will stay there until it is moved back to the center position to turn the saw off. It will be in an on position to the left and right of the center; which is off.

7.1-8. ELECTRICAL COMPONENTS CONNECTING WIRING-The Electrical Components Connecting Wiring is made to specifications of the Switches, Motor, and Battery Plug In Terminals of which it will be installed and connected to. The 120 volt Model-X2 will have the Plug in Terminals without the holes that the Battery will need to make the connection. It will have the screw threaded holes one for each to connect the wiring.

7.1-10. THE BLADE STABILITY GEARS-There are four 4 of these gears that are placed evenly surrounding the blade to keep it in perfect alignment while it is rotating. And oil could be added in this area when a where need to keep the part lubricated.

Tropical Saws & Export Corporation ™

Perimeter Saw-Design Patent Application
Joe Nathan Brown-Inventor, Designer, Author, Engineering

PARTS LISTINGS-and Descriptions
**7. This is Fig. VII from page 29, of the Claims I-1. & Specifications
These are the CLAIMS AND SPECIFICATIONS I-1.**

7.1-11 FRONT COMPONENTS INSERT ATTACHMENT- The Front Components Insert Attachment, is the part that is made to mount the Component Center Cover onto. It is designed to leave an opening in the front of this part of the frame making room for the part that may need to be installed in this area.

7.1-12 THE BLADE STABILITY GEARS-There are four 4 of these gears that are placed evenly surrounding the blade to keep it in perfect alignment while it is rotating. And oil could be added in this area when a where need to keep the part lubricated.

7.1.-13 THE 90 DEGREE ELBOW GEAR-The 90 Degree Elbow Gear is place and installed just below the blade, in the precise location for the maximum contact with the blades' Leverage Activators on the circumference of them. It is comprised of two shafts on the inside; which are connected together by gear at the base of them. And the shafts are put in an engineered frame; which has 4 screw holes on the bracket area base for mounting it to the frame.

7.1-14 UTILITY HANDLE- The Utility & Storage Handle will have the sizing to fit the hands. It will be made in small, medium and large with a connection space of 1" 2" 3" 4" and 5" inches between the gripping pad in the middle of it to the screw attaching holes in the frame. It is attached and mounted to the upper and lower Components Housing Insert Attachment with 4 screws per section. It also will be counter sunk into the two sections for a recessed finish.

7.1-15 ELECTRIC, GAS, OR AIR COMPRESSOR MOTOR-The Electric Motor needed for this Model-X-1, will increase in size by the size of the blades outer diameter; which causes the frame to be larger. Made in 12 sizes and the beginning horsepower are: 1/8hp, 3/16hp, 1/4hp, 5/8hp, 3/4hp, 7/8hp, 1hp, 1 &1/8hp, 1 & 3/16hp, 1 & 1/4hp, 1 & 5/8hp, 1 & 3/4hp also 2hp, 3hp, 4hp, and 5 to 10 horsepower for the larger Heavy equipment Models.

Tropical Saws & Export Corporation ™

Perimeter Saw-Design Patent Application
Joe Nathan Brown-Inventor, Designer, Author, Engineering

These are the CLAIMS AND SPECIFICATIONS I-1.

7. This is Fig. VII from page 29, of the Claims I-1. & Specifications

PARTS LISTINGS-and Descriptions

7.1-16 BATTERY CONNECTION PLUG IN TERMINALS- The Battery Connection Plug-In Terminals is a part made of metal in copper. It has three holes in it, #1 one is in the top rectangular in shape, and this hole allows the maximum contact by the metal socket extended out from the battery end. Whole #2 is just below #1 and is threaded so the screw that holds the Electrical Wiring Connecting Terminals can be screw tightened to it. Hole #3 is at the base of the part an is large enough to allow a screw to slide through and be tightened to the frame.

7.1-17 COMPONENTS INSERT ATTACHMENT, CENTER FRONT COVER-Is located on the front of the saw and is attached and mounted to the Components Housing Insert Attachment, Front Frame. It will be the part that is comprised of the Blade Holding Apparatus in it top, upper section. The Blade Holding Apparatus fits on the front of the blade and is the last piece; which is touched, by material being placed between the blades' cutting edges.

7.1-18 COMPONENT MOUNTING SCREWS- These screws total 60 and will have Phillips head and align wrench head for two possible tightening possibilities. The screws will be surface mounted and also counter sunk into the frame section and components, when and where specified.

7.1-19 BOTTOM COMPONENT COVER- Is located at the bottom of the saw and is attached, mounted to the frame by 2 to 4 screws. It holds the battery in place and will have a hole in the center for the rubber insulation surrounding the 120 Volt Electric Cords.

Tropical Saws & Export Corporation ™

Perimeter Saw-Design Patent Application
Joe Nathan Brown-Inventor, Designer, Author, Engineering

Parts Descriptions-Blade Sizes
Dimensions of the **Outer Diameter** of all Blades is: 1" to 48 "Inches

These are the CLAIMS AND SPECIFICATIONS I-1.
LENGTH

The **4"** Outer Diameter
1. Inner Diameter of: 1/4", 3/8", 1/2", 5/8", 3/4", 7/8"
2. Inner Diameter of: 1", 1" 1/4", 1 3/8", 1 1/2", 1 5/8", 1 3/4", 1 7/8"
3. Inner Diameter of: 2", 2 1/4", 2 3/8", 2 1/2", 2 5/8", 2 1/4", 2 7/8"
The **8"** Outer Diameter
4. Inner Diameter of: 4", 4 1/4", 4 3/8", 4 1/2", 4 5/8", 4 3/4", 4 7/8"
5. Inner Diameter of: 5" 5 1/4", 5 3/8", 5 1/2", 5 5/8", 5 ¾", 5 7/8"
6. Inner Diameter of: 6", 6 1/4", 6 3/8", 6 1/2", 6 5/8", 6 3/4" 6 7/8"

Tropical Saws & Export Corporation ™

Perimeter Saw-Design Patent Application
Joe Nathan Brown-Inventor, Designer, Author, Engineering

Parts Description-Blade Sizes
This is the dimensions of the Outer Diameter of all Blades is: 1" to 48"inch in length.

These are the CLAIMS AND SPECIFICATIONS I-1.

LENGTH

The **12"** Outer Diameter
7. Inner Diameter of: 8", 8 1/4", 8 3/8", 8 1/2", 8 5/8", 8 3/4", 8 7/8"
8. Inner Diameter of: 9", 9 1/4", 9 3/8", 9 1/2", 9 5/8", 9 3/4", 9 7/8"
9. Inner Diameter of: 10", 10 1/4", 10 3/8", 10 1/2", 10 5/8", 10 3/4" 10 7/8"
The **16"** Outer Diameter
10.   Inner Diameter of: 12", 12 1/4", 12 3/8", 12 1/2", 12 5/8", 12 3/4", 12 7/8"
11.   Inner Diameter of: 13", 13 1/4", 13 3/8", 13 1/2", 13 5/8", 13 3/4", 13 7/8"
12. Inner Diameter of: 14", 14 1/4", 14 3/8", 14 1/2", 14 5/8", 14 3/4" 14 7/8"

Page 133. of 140.

Tropical Saws & Export Corporation ™

Perimeter Saw-Design Patent Application
Joe Nathan Brown-Inventor, Designer, Author, Engineering

Parts Description-Blade Sizes
Dimensions of the Outer Diameter of all Blades is: 1" to 48 "Inches

These are the CLAIMS AND SPECIFICATIONS I-1.
LENGTH

The **20"** Outer Diameter
13. Inner Diameter of: 16", 16 1/4", 16 3/8", 16 1/2", 16 5/8", 16 3/4" , 16 7 /8"
14. Inner Diameter of: 17", 17 1/4", 17 3/8", 17 1/2", 17 5/8", 17 3/4", 17 7/8"
15. Inner Diameter of: 18", 18 1/4", 18 3/8", 18 1/2", 18 5/8", 18 3 /4", 18 7/8"
The **24"** Outer Diameter
16. Inner Diameter of: 20", 20 1/4", 20 3/8", 20 1/2", 20 5/8", 20 3/4", 20 7/8"
17. Inner Diameter of: 21", 21 1/4", 21 3/8", 21 1/2", 21 5/8", 21 3/4", 21 7/8"
18. Inner Diameter of: 22", 22 1/4", 22 3/8", 22 1/2", 22 5/8", 22 3/4", 22 7/8"

Page 134. of 140.

Tropical Saws & Export Corporation ™

Perimeter Saw-Design Patent Application
Joe Nathan Brown-Inventor, Designer, Author, Engineering

Parts Description-Blade Sizes
The Dimensions of the Outer Diameter of all Blades is: 1" to 48 "Inches.

These are the CLAIMS AND SPECIFICATIONS I-1.
LENGTH

The **28"** Outer Diameter
19. Inner Diameter of: 24", 24 1/4", 24 3/8", 24 1/2", 24 5/8", 24 3/4", 24   7/8"
20. Inner Diameter of: 25", 25 1/4", 25 3/8", 25 1/2", 25 5/8", 25 3/4", 25   7/8"
21. Inner Diameter of: 26", 26 1/4", 26 3/8", 26 1/2", 26 5/8", 26 3/4", 26 7/8"
The **32"** Outer Diameter
22.   Inner Diameter of: 28",   28 1/4", 28 3/8", 28 1/2", 28 5/8", 28 3/4", 28" 7/8"
23.   Inner Diameter of: 29",   29 1/4", 29 3/8", 29 1/2", 29 5/8", 29 3/4", 29   7/8"
24.   Inner Diameter of: 30, 30 1/4", 30 3/8", 30 1/2", 30 5/8", 30 3/4", 7/8"

Tropical Saws & Export Corporation ™

Perimeter Saw-Design Patent Application
Joe Nathan Brown-Inventor, Designer, Author, Engineering

Parts Description-Blade Sizes
Dimensions of the Outer Diameter of all Blades is: 1" to 48 "Inches

These are the CLAIMS AND SPECIFICATIONS I-1.
LENGTH

The **36"** Outer Diameter
25. Inner Diameter of: 32", 32 1/4", 32 3/8", 32 1/2", 32   5/8", 32 3/4", 32 7/8"
26. Inner Diameter of: 33", 33 1/4", 33 3/8", 33 1/2", 33 5/8", 33 3/4", 33 3/4"
27. Inner Diameter of: 34", 34 1/4", 34 3/8", 34 1/2", 34 5/8", 34 3/4", 34 7/8"
The **40"** outer Diameter
28. Inner Diameter of: 36", 36 1/4", 36 3/8", 36   1/2", 36 5/8", 36 3/4", 36 7/8"
29. Inner Diameter of: 37", 37 1/4", 37 3/8", 37 1/2", 37 5/8", 37 3/4", 37   3/4", 37 7/8"
30. Inner Diameter of: 38", 38 1/4", 38 3/8", 38 1/2", 38   5/8", 38 3/4" 7/8"

Page 136. of 140.

Tropical Saws & Export Corporation ™

Perimeter Saw-Design Patent Application
Joe Nathan Brown-Inventor, Designer, Author, Engineering

Parts Description-Blade Sizes
Dimensions of the Outer Diameter of all Blades is: 1" to 48 "Inches

These are the CLAIMS AND SPECIFICATIONS I-1.
LENGTH

The **44"** Outer Diameter
31. Inner Diameter of: 40", 40 1/4", 40 3/8", 40 1/2", 40 5/8", 40 3/4", 40 7/8"
32. Inner Diameter of: 41", 41 1/4", 41 3/8", 41 1/2", 41 5/8", 41 3/4", 41 7/8"
33. Inner Diameter of: 42", 42 1/4", 42 3/8", 42 1/2", 42 5/8", 42 3/4", 42 7/8"
The **48"** Outer Diameter
34. Inner Diameter of: 44" 44 1/4", 44 3/8", 44 1/2", 44 5/8", 44 3/4", 44 7/8"
35. Inner Diameter of: 45", 45 1/4", 45 3/8", 45 1/2", 45 5/8", 45 3/4", 45 7/8"
36. Inner Diameter of: 46"46 1/4", 46 3/8", 46 1/2", 46 5/8", 46 3/4", 46 7/8"

Tropical Saws & Export Corporation ™

Perimeter Saw-Design Patent Application
Joe Nathan Brown-Inventor, Designer, Author, Engineering

Parts Description-Blade Sizes

THE **DIMENSIONS** OF THE **WIDTH & THICKNESS** of the Blade
For the 1" to 48" Outer Diameter

These are the CLAIMS AND SPECIFICATIONS I-1.

1.	4"	1/32", 1/16", 1/8"
2.	8"	1/32", 1/16", 1/8"
3.	12"	1/16", 1/8", 3/16", 1/4"
4.	16"	1/16", 1/8", 3/16", 1/4"
5.	20"	1/16", 1/8", 3/16", 1/4"
6.	24"	1/16", 1/8", 3/16", 1/4", 3/8", 1/2"
7.	28"	1/16", 1/8", 3/16", 1/4", 3/8", 1/2"
8.	32"	1/16", 1/8", 3/16", 1/4", 3/8", 1/2"
9.	36"	1/16", 1/8", 3/16", 1/4", 3/8", 1/2"
10.	40"	1/16", 1/8", 3/16", 1/4", 3/8", 1/2"
11.	44"	1/16", 1/8", 3/16", 1/4", 3/8", 1/2"
12.	48"	1/16", 1/8", 3/16", 1/4", 3/8", 1/2"

Tropical Saws & Export Corporation ™

Perimeter Saw-Design Patent Application
Joe Nathan Brown-Inventor, Designer, Author, Engineering

Appendix A-1
The Actions Taken For the Documentation:
To Protect and Secure the Intellectual Property Rights;

1. The Corporate Documentation, information is... To file and receive a receipt and the date of mailing, has taken place in that , Tropical Saws & Export Corporation is registered with the State of Florida, under Corporation. Document Number: P08000108343

2. The Patent Documentation, Provisional Patent, information: To file and receive a receipt and the date of mailing, has taken place in that the Perimeter Saw, has been filed and receive for Patent Pending Status.
Filing Date: 4-14-2004

3. The Patent Documentation, Disclosure Document Deposit Request, Information.... To file and receive a receipt and the date of mailing
Filing Date: 10-10-2003

4. This is the Design Patent Application, information.... To file and receive a receipt and the date of mailing & on line filing, The Non-Provisional Patent has been filled as of:
Filing date: 3-16-2009

5. This is the Design Patent Express Mail Label, information....
To file and receive a receipt and the date of mailing & on line filing
Mailing date: 3-12-2009

6. This is the Copyright Application, Form VA, Visual Arts information....
To file and receive a receipt and the date of mailing & on line filing

Mailing date: 3-17-2009 with the use of expresses Mail Label Number: EH-602931969 US E on Line-filing Date: 5-22-2009 and 5-27-2009

7. The Copyright Application, Form TX. Text Information....
To file and receive a receipt and the date of mailing & on line filing
Mailing Date: 3-17-2009

Page 139. of 140.

Tropical Saws & Export Corporation ™

Perimeter Saw-Design Patent Application
Joe Nathan Brown-Inventor, Designer, Author, Engineering

The Actions Taken For the Documentation:
To Protect and Secure the Intellectual Property Rights;

Appendix A-2
8. The Book in Conclusion: Dear Readers, the inventor would like to thank you
for the interest you have shared in purchasing the materials herein known by the
title of: Tropical Saws & Export Corporation, Perimeter Saw, Design Patent
Application Claims & Specifications.

The next book will be begin on December 28, 2010 and will be entitled: Tropical Saws &
Export Corporation, Perimeter Saw, Design and **Utility** Patent Application Claims &
Specifications.

The contact information just below the title on the cover page allows you a personal and
business opportunity to reference this tool and other products currently being developed
by the company. So, please address all communication to Mr. Joe Nathan Brown, President
& Chief Executive Officer of Tropical Saws & Export Corporation. The opening for
Employment, Contracts and Joint Venture participation is now available, as December 23,
2010.

The e mediate opening is in sales of the new book publication that will be sold at book
stores and on INTERNET location. If you are a publisher in the need of new materials for
publication, please fill free to contact us: by E-mail etc.

This book highlights the work done to make a product from the starting date with the
sketches and drawings to the completion date, as the Perimeter is to be made and or
manufactured. It will give you suggestions on documentation of your work as it relates to
the work being done on the **Perimeter Saw,** new product.

The Marketing and Advertising will reflect fields in; which the tool will be used and
distributed. Some of the Employment fields that will benefit with the purchase of this book
and finally the new tool when it is built are:

1. **Construction-** Electrical, Plumbing, Pipe Fitting, Steel & Iron Workers
2. **Construction-** Oil Fields Drilling and Platform Building
3. **Manufacturing-** Tubing & Pipe Fabrication, Nut, Bolts & Nail Fabrication
4. **Manufacturing-** Ship building, Auto, Car and Trucks, Motorcycles
5. **Farming-** Agriculture, Gardening, Landscaping, Lawn Maintenance Services

The Book on the Perimeter Saw will be sold at a Retail price of: $9.99.
The Perimeter Saw when made and completed will retail at hardware stores, department
stores and tool supply outlets at a price estimation of **$149.00 to $1,249.00** US Dollars.

Page 140. of 140 .

Tropical Saws & Export Corporation ™

www.ingramcontent.com/pod-product-compliance
Lightning Source LLC
Chambersburg PA
CBHW021955170526
45157CB00003B/993